ALSO BY AMIR D. ACZEL

Probability I:
Why There Must Be
Intelligent Life in the Universe

God's Equation:
Einstein, Relativity,
and the Expanding Universe

Fermat's Last Theorem:
Unlocking the Secret of
an Ancient Mathematical Problem

The Mystery of the Aleph:
Mathematics, Kabbalah,
and the Search for Infinity

The RIDDLE of the COMPASS

The Invention That Changed the World

AMIR D. ACZEL

HARCOURT, INC.

New York   San Diego   London

www.harcourt.com

Line illustrations by Patti Isaacs, Parrot Graphics

Library of Congress Cataloging-in-Publication Data
Aczel, Amir D.
The riddle of the compass/Amir D. Aczel.—1st ed.
p.   cm.
Includes bibliographical references and index.
ISBN 0-15-100506-0
1. Compass.   I. Title.
VK577.A29 2001
912'.0284—dc21   00-047153

Text set in Dante
Designed by Kaelin Chappell

First edition

A C E G I K J H F D B

Printed in the United States of America

For Debra

Contents

Preface ix

ONE Odyssey *1*

TWO Signs in the Sea and Sky *9*

THREE Dante *29*

FOUR The Etruscan Chandelier *39*

FIVE Amalfi *53*

SIX The Ghost of Flavio Gioia *63*

SEVEN Iron Fish, Lodestone Turtle *77*

EIGHT Venice *91*

NINE Marco Polo *111*

TEN Charting the Mediterranean *123*

ELEVEN A Nautical Revolution *133*

TWELVE Conclusion *153*

A Note on the Sources 161

References 165

Acknowledgments 169

Index 171

Preface

THE LATTER PART OF THE THIRTEENTH CENTURY
marked a new beginning in world history. If the
twentieth century was the time of the information
revolution and the eighteenth century was the start of the
industrial revolution, then the end of the thirteenth century
could rightly be called the beginning of the commercial
revolution.

Within a few decades of 1280, the world saw a dramatic
rise in trade, and with it, increased prosperity for maritime

powers such as Venice, Spain, and Britain. A single invention—the magnetic compass—made this possible. The
compass was the first instrument that allowed navigators, at
sea, on land, and—much later—in the air, to determine
their direction quickly and accurately at any time of the day
or night and under almost any conditions. This allowed
goods to be transported efficiently and reliably across the
seas and opened up the world to maritime exploration. The
earth would never be seen the same way again.

The compass was therefore the most important technological invention since the wheel. With the exception of
ancient weighing scales, the compass was also the first me-
chanical measuring device ever invented, as well as the first
instrument with a pointer, allowing a person to visualize a
measurement—in this case a direction.

The importance of the magnetic compass cannot be
overestimated. Today, seven hundred years after the emergence of the compass with a compass card indicating directions, and a millennium or longer since the invention of the
simpler needle compass, every ship carries a magnetic compass at least as a backup for its electronic instruments.

But the magnetic compass was not only a celebrated technological and scientific invention. It also became a metaphor in poetry and has long been a device in mystical
investigations and divinations. Since the dawn of civilization, people have been fascinated with the natural phenome-

non of magnetism. The lodestone, because of its puzzling ability to exert force at a distance on metallic objects, was believed to possess mysterious, supernatural properties. Centuries before the compass was known in the West, Chinese diviners used a magnetic compass to help make decisions and prophecies. In Europe, especially around the Mediterranean Basin, cults that used magnetic devices also flourished.

The origins of the compass are shrouded in mystery. Or rather, the story of the compass is a series of mysteries which have not, until now, been satisfactorily addressed. The tale of the invention of the magnetic compass spans the breadth of human civilization. Geographically, the story traverses the world, from China to the Mediterranean, Scandinavia, Arabia, Africa, and the New World. As a history, the story covers events that took place in ancient times, during the medieval era, and that continue to our own time. This book explores the series of riddles that make up the story of the compass—the mysteries of the invention that changed navigation, commerce, and the world economy.

The compass works because the earth is a giant magnet. A magnet is an object that induces a magnetic field; this field is a region of space surrounding the magnet within which invisible lines of force exist that run between two points called the magnet's north and south poles. A magnetic field is induced every time electrons move, such as when an electric current flows. Natural magnets, such as the lodestone, derive their magnetism from the peculiar way in which the electrons move within them. The magnetic field exerts a pulling action on iron and similar elements and either attracts or repels other magnets, depending on their orientation. Like poles repel; unlike poles attract. If a magnet is brought into a magnetic field, assuming it can move freely, it will align itself with the new magnetic field.

The earth's core of molten iron swirls around in a spherical pattern, deep under the surface. The currents generated as this mass of liquid iron rotates below the earth's crust create a dynamo action whose result is magnetism. These currents turn the entire earth into a giant magnet with a magnetic field and a north and a south pole.

The compass needle is a small magnet suspended in air or water so it can rotate freely and orient itself. This magnet reacts to the magnetic field produced by the huge magnet, the earth, and aligns itself accordingly. This is demonstrated in the illustration on the facing page.

Earth's magnetic north pole was not always in the direction we know as north, and Earth's magnetic south pole was not always to the south. The earth's polarity will remain constant for hundreds of thousands of years, and then suddenly switch. In such a switch, the north magnetic pole becomes the south magnetic pole, and vice versa. Scientists

Preface

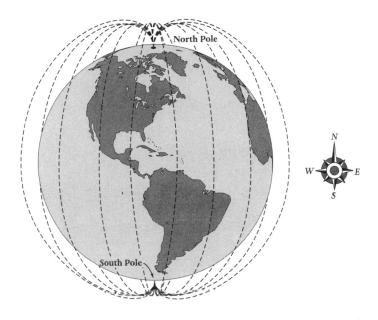

How the magnetic compass works

have deduced the existence of this mysterious phenomenon by studying geological sediments of elements that align themselves with the magnetic field of the earth and estimating the time when these elements were free to move and orient themselves before solidifying in place. We don't know what causes the switches in the polarity of the planet or when the next reversal might take place. But if you sailed a ship 300,000 years ago, before the last switch in the earth's polarity, your compass needle would have pointed south instead of north.

Preface

The compass is a reliable instrument, but some factors affect its performance. The earth's magnetic north pole deviates somewhat from the earth's true geographic North Pole. This deviation from true north changes from place to place and over time. Aided by scientific tables and charts, however, a navigator can correct the relatively small error introduced by the difference between magnetic north and geographic north and navigate accurately. Metal objects aboard ship may also cause a compass to deviate from Earth's magnetic north. This deviation can be corrected by adjusting the compass to its surroundings, most often by the placement of two large metal balls on its sides. Once these corrections are made, the magnetic compass is a very reliable tool in navigation.

How did people discover that a magnetic needle, suspended in air or water, could be used to indicate the north? Where did the idea of north, south, east, and west originate, and how did mariners learn to use these directions? How did sailors navigate the seas before the advent of the compass? And how did they begin to use the compass for navigation? These are some of the riddles we will attempt to solve in this book.

The story of the compass is a great tale of human ingenuity. It is a story of invention, innovation, opportunity, and capitalism. It is an account of how one civilization made an important invention and how another civilization on the other side of the world put that invention to use, promoting trade and creating wealth. The story of the compass is the story of human civilization and its ability to flourish and prosper by invention and opportunity, by developing a technology and exploiting its promise.

The RIDDLE of the COMPASS

ONE

Odyssey

I FIRST BECAME INTERESTED IN THE COMPASS WHEN I was a child. I grew up on a passenger ship in the Mediterranean; my father was the ship's captain. I spent all of my childhood years aboard ship, with the exception of the few months each year I stayed ashore attending school. When out to sea, I made up missed work by correspondence with my teachers. Somehow, I managed to complete my education this way. But on board the ship I learned other things.

When I was ten, my father taught me how to steer a ship.

A crew member brought me a little stool so I could reach the wheel. First, my father held the wheel with me; then I learned to steer by myself, following my captain's commands. "Port ten," my father would order. "Port ten, sir!" I'd respond, and steer. "Starboard five," he would call. "Starboard five," I'd repeat, and turn the wheel. Then came the harder work: "Steady as she goes," my father would order, and I would have to use the ship's compass to keep the course precisely to the one shown at the moment my father issued the command. This was difficult—a ship is not a car (as I'd learn years later when I reached legal driving age). A ship responds slowly and has inertia. Even when the rudder points straight, the ship will continue to turn if it has already started the motion, so to stop the turn, you have to steer the other way until the ship responds. Then you have to turn the wheel in the opposite direction in anticipation of stopping the ship's turn at the point you need to have it steady. Steering by compass is both an art and a science, as I learned at the age of ten.

As the years went by, I developed a feel for the compass and the wheel. Appreciating my father's confidence in me—this was a ship with seven hundred people on board—I worked hard to excel. Many years later, I can still hear in my mind the ticking of the compass, degree by degree, as the ship turned, the pace of the ticking reflecting the swiftness of the turn and hence the severity of the move I'd have to make if I needed to stop the motion. The greatest challenge I faced as a young helmsman took place only three or four years after I'd first taken the wheel, when my father asked me to steer his ship through the Strait of Messina.

The Strait of Messina is a narrow sea passage between the region of Calabria, at the southern end of Italy, and the island of Sicily. Because the strait connects two seas within the Mediterranean—the Tyrrhenian and the Ionian—which are separated by large pieces of land, fierce currents build up from the southern entrance of the strait to its narrow ending by the Sicilian city of Messina. Steering a ship through the strait is a challenge even for the most experienced helmsman. It was night, and I could see the distant lights of villages and towns on both sides of the ship as we worked our way north. When we neared the bottleneck at Messina, the ship began to shake, roiled by the terrific currents that converge at the narrows. As the currents built up in intensity, the ticking of the compass gained momentum, and I had to steer port, then quickly starboard, then port faster to prevent oversteering. At times it seemed the ship would succumb to the currents, but I refused to let the sea win this battle. As we exited the strait and again found ourselves in calm seas, my father came over. We stood together for a moment, watching the amber glow of lava spewing into the air at regular intervals on the distant island of Stromboli in the Tyrrhenian Sea, which we had just entered. "You did well," he said quietly. We were safely on our way to our destination in southern Italy.

Many years later, I again found myself in sunny, hospitable southern Italy. And again, a compass guided me there. I had come in search of the origins of the mysterious device that has so fascinated me since childhood—the instrument that revolutionized navigation.

As soon as I left Salerno and drove west along the coast toward Amalfi, the road became extremely curvy. I had to downshift, but the Alfa Romeo 156 was made for such treacherous driving—the engine whirred and the wheels hugged the pavement, not giving an inch, as I followed the first hairpin turn. It was a Friday afternoon in early summer, and too many drivers were braving the narrow road on the precipitous cliffs.

I glanced around me. On my right, a sheer rock face rose skyward; on my left, the cliffs plummeted to the sea. As I neared my destination, the vegetation became more dense: olive trees with gnarled trunks; red and white oleanders; purple bougainvilleas; wild lemon and orange trees, their branches heavy with ripe fruit. A few miles farther, I started to see the stone and stucco houses the people of the Costiera Amalfitana have built into the rocky slopes. An hour later, as the car made one last sharp turn and emerged from a short tunnel, I saw below me, by a deep blue bay, the town of Amalfi. I parked on the side of the road and walked down the narrow stairs to the old harbor, passing well-kept houses with geraniums in window boxes. On the way down I came across a hotel with a faded sign: HOTEL LA BUSSOLA ("The Compass").

Soon I found myself in the center of Amalfi, a town situated by a small harbor. Above an archway, I saw a bronze plaque with an inscription in Italian. Translated, it read:

All of Italy, and Amalfi, must give credit to the great invention of the magnetic compass, without which America and other unexplored places would not have been opened to civilization. Amalfi commemorates this pure Italian glory with special honors to its immortal son, Flavio Gioia, the fortunate inventor of the magnetic compass.—1302–1902.

Near the green town square, a small obelisk bore a plaque dated 1902 with the chiseled inscription: AMALFI TO FLAVIO GIOIA, INVENTOR OF THE COMPASS. Across the street from it, facing the Mediterranean, was a tall bronze statue of a hooded man, looking down at an instrument in his hand. He looked like a cross between Dante and Columbus, perhaps not by chance. A simple plate at the foot of the statue bore a cross and a name: FLAVIO GIOIA.

Historical sources I had consulted on the compass named

Amalfi as the place of its European invention, with some references mentioning the name of Flavio Gioia. On the streets of Amalfi and on every historical marker, he was very much alive—but who was he?

I walked to the bookstore in the main square. There were books on every subject, in Italian and other languages. But there was not a single book or pamphlet on the life of Amalfi's most illustrious son, not a word about Flavio Gioia. I asked about Flavio Gioia on the streets, in stores, and at the tourist center, but nobody seemed to know where I might find information about the man and his invention. I walked past the bus stop. The sign for the bus proclaimed the name of the local bus company: FLAVIO GIOIA. In Amalfi, Flavio Gioia was at once everywhere and nowhere to be found. I was determined to find out more about the elusive inventor of the compass, but where? Finally, a policeman tipped me off.

"Try the Amalfi cultural center," he said in response to my question about Flavio Gioia. He directed me to a location in a back alley, away from the town center with its sunseeking vacationers. I walked the narrow streets of the hidden part of Amalfi, climbed a set of stairs, and turned around an architecturally undistinguished building to enter the center. "Oh yes, we do have some material on Flavio Gioia," said the archivist. "But, you know, it isn't at all clear that the man ever existed. Here, read this before you go any further," he said, as he handed me a pamphlet quoting the words of the Italian historian Padre Timoteo Bertelli. I began to read:

> *Flavio Gioia never existed. He represents only a kind of myth, created late after his presumed lifetime, and hence suspect. He is a fantasy produced by the fertile southern imagination of the people of Amalfi and elsewhere. . . .*

"So this is what I came all the way here for . . . ," I muttered. "The fertile southern imagination?" I looked up from Bertelli's writings and saw the archivist's gentle smile. His eyes reflected the wisdom of generations of Italian scholars, archivists, editors, careful collectors of ancient facts.

"Don't despair so quickly, Professore," he said. "You came a long way, but I think you have arrived at the right place to solve your riddle."

There was a thud as he dropped a pile of dusty old volumes in front of me, and he quickly excused himself and disappeared into his office.

I sat there, in the sweltering reading room of the Center for the Culture and History of Amalfi, and took down the top volume from the stack. I opened its yellowing pages and began to read a curious book: a two-hundred-year-old treatise written in French but published in Naples. Its author had made a thorough study of ancient navigation and claimed to have identified the methods of navigation used by Odysseus.

TWO

Signs in the Sea and Sky

HOW DID NAVIGATORS OF ANTIQUITY FIND THEIR
way at sea in the days before the compass? There is a
myth, propagated by people with little understanding of the
sea and no faith in human ingenuity, that ancient mariners
navigated by hugging the coastline. Nothing is further from
the truth. Since time immemorial, mariners have sailed
across seas far from the sight of land, and early sailors who
inspired the stories of the Bible and Greek mythology were
quite adept at navigating in the open seas without the

advantages of the compass. Recently scientists reported the finding of a 2,300-year-old shipwreck in the middle of the Mediterranean Sea, 200 miles from any shore, confirming the assertion that ancient mariners did not hug the coast.

The Minoan civilization of Crete, an island in the middle of the eastern Mediterranean, was an ancient maritime empire whose wealth derived from extensive trade with other nations. To sail anywhere from Crete, one must cross open waters and remain far from any coast for at least some time. The Cretans navigated across the Mediterranean with great success. In fact, their major trading partner was Egypt, over three hundred miles of open sea to the southeast. Bronze Age frescoes (dated to 1600 B.C.) found on Crete and at the Minoan site of Akrotiri on the neighboring island of Santorini show relatively large ships with sails and oars. These ships traversed the seas between Minoan harbors and faraway lands. Minoan mariners regularly crossed the eastern Mediterranean, spending days and weeks out of sight of land.

By all accounts, the Phoenicians and the ancient Israelites were seafaring peoples as well. There is copious evidence that these early navigators did not hug the coasts. When Jonah's ship encountered a storm at sea and could not reach land, he was tossed overboard and swallowed by a whale.

King Solomon traded with the mythical Ophir across the seas and courted the Queen of Sheba.

The ancient Egyptian hieroglyphic for a foreign ship featured a square sail, while Egyptian ships were symbolized by different sail designs. From records found at archaeological sites in Egypt, we can deduce that ships of many nations regularly arrived in Egypt, even during very early times.

Roman records of navigation give directions for sailing across the sea—for example, from the Greek islands to Egypt. Saint Paul's shipwreck on the island of Malta is vividly described by an eyewitness. Malta lies in the middle of the Mediterranean Sea, between Sicily and the North African coast. The scriptural record describes how the sailors lost hope after clouds covered the sky for many days and the sun and the stars were not visible, so that they could not navigate.

There is evidence that the Hawaiian Islands were first inhabited fifteen hundred years ago by Polynesians who sailed there in large canoes from the Marquesas Islands, thousands of miles of open ocean away.

The belief that before the compass navigators "hugged the coastline" could only be held by people who do not understand ships or navigation. The greatest danger a mariner faces is that of running aground. The reason for this is that the depth of the sea varies widely, and a navigator cannot always predict where the sea might not be deep enough for the ship's draft. Rocks and shoals abound as well, often miles from shore. With all these hazards along the coastline, a sea-faring person is safest in open waters. Ancient mariners sailed wherever they needed to, and even if they had to go from one location to another on the same coast, they did not hug the coastline but stayed a safe distance out at sea.

The danger of hidden rocks and sandbanks helped create the navigator's first instrument, the sounding line. This was a simple implement: a long line with knots marking distance along its length, and a lead weight at its end. The sounding line was considered such an important tool in early navigation that ships detained in port for nonpayment of duty or for other reasons had their sounding lines confiscated. This custom prevailed even through periods when ships carried a variety of other, more modern instruments. Samuel Clemens's choice of the pen name Mark Twain was based on the practice of using the sounding line on ships going down the Mississippi River in the nineteenth century.

The sounding line was used to tell the navigator the depth of the sea, when such depth was within the range of the line. The bottom part of the piece of lead at the end of the line was often rubbed with tallow so that when the sounding line was brought up, the mariner could see what type of sediment lay at the bottom of the sea. Navigators learned to recognize the type and color of the sand or silt or seaweed brought up by the line, and this information was used in navigation.

The bottom of the sea shows a pattern of crests and troughs, elevated areas and ravines. Knowledge of these patterns is important in navigation, and before the use of charts with depth markings, navigation relied on the captain's or pilot's knowledge of the ocean floor. As a ship approached a shore, repeated soundings were taken to enable the mariner to assess the rate of decrease in depth and avoid hitting the bottom. This practice dates from biblical times. When the navigators of Saint Paul's ship couldn't see the sky for many days, they lost their bearings. One night at midnight, they suspected they were near shore and cast the lead. The first sounding was twenty fathoms, the next one fifteen, so they knew that they were fast approaching shore. They ran the ship aground, seemingly deliberately, the following day, on the shore of what the Maltese now call Saint Paul's Bay.

Along with a knowledge of the seafloor, a navigator had an understanding of the tides, learned from experience. This was more consequential in the Atlantic and Indian Oceans than in the Mediterranean, where tidal changes are small. Since early antiquity, as ships approached port, they hired local fishermen to help their captains navigate into the harbor. These fishermen, who had excellent knowledge of the tides and the bottom of the sea at their location, were the precursors of the professional pilots that shipping companies hire to this day to aid their captains when entering and exiting ports.

Once a ship was within sight of land, knowledge of the shore profile was of paramount importance to the ship's navigator. Before charts and other aids were introduced, mariners relied on their memory and experience to identify the location of their destination port. Promontories, inlets, capes, and bays could be identified from relatively far out at sea. Capes, especially, were useful aids to navigation because of their prominence, but they also presented a challenge to the navigator. In coastal sailing, capes are difficult for a ship to round because of the prevalence of storms and unpredictable gusts. Near capes, the currents can be treacherous as well. For these reasons, navigators preferred to stay out at sea rather than hug the coastline and hazard rounding a cape too close to shore.

As Homer tells us in the *Odyssey*, King Menelaus, sailing home from Troy, experienced his first misfortune at Cape Sunion, where Attica juts into the Aegean Sea. Here his helmsman, the world's best steerman in a gale, was struck dead. When the fleet continued south after burying the helmsman, it approached Cape Malea—a southern point of the Peloponnisos notorious for foul weather. Here Zeus

sent the fleet a howling gale with waves as high as mountains. The fleet was scattered, and Menelaus was driven
across the sea to Egypt, while the rest of his fleet was lost on
the shores of Crete.

Since the dawn of civilization, lighthouses have been built
on capes and high terrain overlooking the sea to help navigators find their way. Cape Malea has had a lighthouse since
time immemorial. Next to the lighthouse there is a small
chapel, and a monk lives there, isolated from the rest of civilization by miles of empty space. Whenever my father's ship
approached Cape Malea, my father would blow the ship's
horn three times in salute. The monk would appear, waving
a flag and ringing the chapel's bell until our ship disappeared
behind the cliffs. The passengers, told in advance, would all
come on deck to watch and wave and participate in this maritime tradition of friendship. When he was a young officer,
my father learned this custom from his captain, who, in
turn, had learned it from his captain, and so on back in time.

One of the seven wonders of the ancient world, the Colossus of Rhodes, was also an aid to navigation. A giant
statue of the Greek god of the sun, Helios, the Colossus
was well over a hundred feet high, the masterpiece of the
renowned sculptor Chares of Lindos. This giant statue
straddled the entrance to Rhodes harbor, and it was so big
that ships in full sail could enter the harbor by passing right

under it. The Colossus was designed to help navigators find the island and the entrance to its harbor.

Other aids to navigation included knowledge of winds, currents, and the habits of various animals. Understanding the winds and currents was immensely important to ancient mariners, since winds and currents tend to follow known patterns, which vary with the season. As we will see, the directions on the compass card got their names from those of prevailing winds.

The migration patterns of birds and the location of particular sea animals also provided clues to the mariner about the ship's location. Sea snakes abounded miles from shore off the coasts of India, so when a navigator saw them he could safely assume he was approaching the shore. Birds were especially useful to navigators of antiquity. The migration routes of various species of birds are constant from year to year, and thus a mariner could follow these birds and decide which course to follow.

In the early centuries of our era, long before navigators began to use the compass, Irish monks traveled from one island to the next in small boats under perpetually cloudy skies. They were able to find their way with confidence by following migratory birds. The Norsemen discovered Iceland in A.D. 870 by following winds and birds. Navigators sometimes took an active role in using birds in navigation by carrying them on board. The Vikings carried ravens with them, and when they thought they might be close to land, they released one of their birds. If the raven flew away, the ship's captain would follow it and likely find land. If the bird returned to the ship, he would deduce that there was no land nearby. This technique goes back to Noah. The Bible tells us that Noah released a dove and that it returned with an olive branch.

Ravens and other birds may sense the shore when they are close to it; this may be how they find their direction at sea. But for migrating over large stretches of open sea, animals must have a sense of direction. After traversing the wide ocean for two years, salmon find the rivers in which they began their lives, and migrating birds can fly over great distances, always finding their way to their destination. How do they do it? In 1997, Michael Walker of Auckland University in New Zealand and his colleagues reported in the journal *Nature* that they had identified the facial nerve fiber in trout that fires in response to a magnetic field. The research group has also studied the magnetic sense of honeybees, yellowfin tuna, sockeye salmon, fin whales, and homing pigeons. In these animals and others, the scientists have found a sensitivity to fluctuations in Earth's magnetic field. It is likely that many birds, reptiles, and mammals can orient themselves relative to Earth's magnetic poles. Walker and his colleagues may thus have identified the actual mechanism that allows animals to find their direction while migrating far from identifiable terrain. It appears that these animals have a built-in compass. Thus navigators of antiquity, long before the invention of the magnetic compass, may have been using a compass borrowed from the animals.

Throughout history, mariners have relied on currents and winds and the shape and depth of the bottom of the sea. They have also observed the habits of birds and sea animals. But the most important navigational guide before the invention of the compass—that is, for the long period from early antiquity to about a thousand years ago—was found not in the water or its vicinity but high in the sky.

Sometime in the third millennium B.C., an astronomer in ancient Egypt looked at the predawn sky and saw the brightest star, Sirius, rising in the east. On the same day, the annual inundation of the Nile began. The Egyptians saw in this coincidence a heavenly sign and built their calendar around it. The first day of the year began with the heliacal—"with the sun"—rising of Sirius, the Dog Star. The calendar devised in ancient Egypt was vastly superior to the lunar calendars of Babylon and early Greece, which required frequent intercalations—insertions of additional days into the calendar in order to correct it.

The Riddle of the Compass

The Egyptian calendar was so successful that Greek astronomers adopted it, and it eventually became the calendar of Western culture. The Egyptians studied the movements of other stars and constellations in the heavens, noting their risings and settings during each day of their year. The Egyptians were the first to divide the day into 24 hours. Our present division of the day into 24 hours of 60 minutes each is the result of a Hellenistic modification of the Egyptian practice, combined with the Babylonian base-60 counting system. The Egyptians recognized thirty-six constellations or individual stars lying on the ecliptic—the arc in the sky marking the path of the sun during the day. (These constellations gave rise to the signs of the zodiac.) Each constellation or star was associated with ten days of the year. Egyptian astronomers kept track of where each constellation was during each hour of each day. On coffin lids of the period from 1800 B.C. to 1200 B.C., we find pictures of the constellations and the associated times of night and day. Thus the ancient calendar was also transformed into a complete star-clock.

Since Egypt was a large country, stretching far from south to north, by traveling from Upper to Middle to Lower Egypt by the Red Sea, ancient Egypt's astronomers were also able to determine early on that the latitude made a difference in how high in the sky a given star or constellation would appear. So while the seasons of the year and the hours of the day determined where stars appeared from east to west, the latitude determined where a star or constellation appeared on the north-south axis. Thus, in principle, by knowing the exact time and date and measuring the positions of stars in the heavens, an astronomer would know his or her location. This principle could be used by a navigator to estimate a ship's position.

The problem was, of course, that in ancient times there were great uncertainties about time, and a scientific mechanism for finding location from the positions of stars would have to wait for many centuries. Approximate calculations could be made, however. The problem of latitude was relatively easy, since it did not require exact knowledge of time. Determining the longitude was a much harder issue, since the mariner had to know the exact time. This problem required the use of accurate clocks and was only solved in the eighteenth century. (See Dava Sobel's book *Longitude* for a description of the problem and its solution.)

Egyptian navigators plying the north-south Red Sea shipping routes could estimate their latitude by observing how high a star climbed in the sky before beginning to set. The latitude in this case was not an exact measurement but a rough idea. In the Northern Hemisphere, the ecliptic lies higher in the sky as one moves south, and the same happens with stars. A navigator could note that Sirius reached a certain elevation in the sky at its maximum (when it crossed the observer's meridian) at the home port and a higher altitude farther south. On the way back, when Sirius's maximum elevation in the sky approached the position it had at the home port, the navigator would know he was getting close to home. Such calculations and measurements, while rough, were very useful in navigation. Because they entailed almost exclusively a north-south sailing, the Red Sea voyages played an important role in improving knowledge of navigation.

Such observations gave navigators of antiquity the knowledge of how to use celestial information to estimate the location of a ship on a north-south axis, and hence an understanding of changes of latitude. The estimation was carried out with the aid of a primitive variant of the sextant, an instrument by which a mariner visually measures the angle of a star above the horizon.

North-south celestial effects were also observable in Mediterranean sailings, although the Mediterranean stretches more widely from east to west. Egyptian mariners are known to have voyaged regularly to Byblos in Syria. This voyage entailed almost exclusively a change in latitude, which could be measured. The Roman historian Pliny wrote about the shape of the earth, describing how stars rise as a navigator moves south:

These phenomena are most clearly discovered by the voyages of those at sea . . . the stars that were hidden behind the curve of the ball suddenly become visible, as if rising out of the sea.

However, the most important use navigators of antiquity made of astronomical observations was not in the inexact estimation of location but rather in the determination of direction.

As the earth rotates about its axis, two points in the sky— the north and the south celestial poles—remain constant as the sky appears to rotate above the observer. During our epoch, a relatively bright star happens to be located very close to the north celestial pole. This star is appropriately named Polaris, or the North Star. To find Polaris, all we have to do is find the Big Dipper and follow its "pointer stars" to Polaris.

Because of the process of precession—the slow rotation of Earth's axis over thousands of years—Polaris was not located at the pole during ancient times. About 800 B.C., Kochab (Beta Ursae Minoris), another star in the same constellation as Polaris, was located relatively close to the pole. Navigators of antiquity knew how to find the Big and the Little Dippers and thus how to identify the North Pole. Once they knew the direction north, all the other directions fell into place. The Greek mathematician Thales, who lived around 600 B.C., wrote that the Phoenician mariners of his time were so adept at navigating by identifying the Little

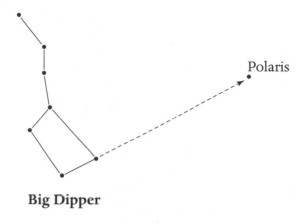

Big Dipper

The Big Dipper and Polaris

Dipper and hence Kochab, that they did not need to use the larger, but more distant, Big Dipper.

Chinese navigators in early times used a jade disk to help them identify the Big and the Little Dippers and, through them, the North Pole. From the location of the stars' positions on one such disk, called the Hsuan-Chi, scientists have been able to date it to 800 B.C. This illustrates how the exact science of modern astronomy helps us establish ancient chronologies. Since we know how the stars move due to the process of precession relative to an observer on Earth and can use this information to compute their positions during early times, we are able to date ancient artifacts depicting the relative positions of stars in antiquity.

During daytime, navigators could determine their

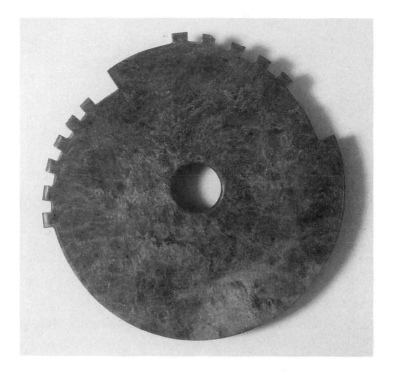

Hsuan-Chi, a Chinese jade disk dated to 800 B.C. and used in navigation to identify the North Celestial Pole. *The National Maritime Museum, Haifa, Israel*

approximate direction by tracing the path of the sun through the sky along the ecliptic, although this method is much less accurate than the nighttime identification of the polestar. The ecliptic changes its location throughout the year, but early navigators were aware of these movements and could tell the direction south by the location of the sun at its highest point in the sky. East and west could be determined, at least roughly, from the rising and setting of the sun.

Odysseus, when his ship was lost, said that east and west meant nothing to him there—being lost meant not knowing the two most basic of directions: the direction of the sunrise and that of the sunset. The midday position of the sun, which in the Northern Hemisphere is south, divides east from west and also gives us, as its direct opposite, the direction north. So the midday position of the sun gave mariners a better estimate of all four directions than did observations during other times of the day. Looking at the sun with unprotected eyes was one of the dangers of navigation in ancient times.

At night, Odysseus steered by a star. When leaving Calypso's island, he "sat and never closed his eyes to sleep, but kept them on the Pleiades, or watched the late-setting Arcturus and the Great Bear. . . . It was this constellation that the wise goddess Calypso had told him to keep on his left hand as he made across the sea." Since the Pleiades and Arcturus differed by nearly eleven hours in right ascension, one or the other would always be visible at night, and Odysseus could keep his bearings.

So day and night, a ship could be guided to its destination with varying degrees of accuracy by observations of the sky. But bad weather posed a serious hindrance to navigation. When the skies were covered with clouds, it was impossible

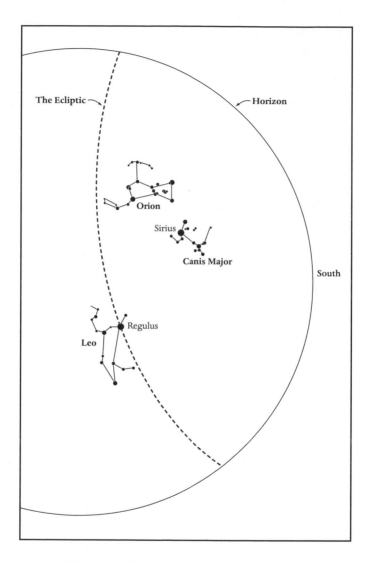

The ecliptic (winter, northern latitudes at night)

for mariners to find where they were going. Mostly for this reason, throughout ancient times, the seas were closed to navigation in winter.

Ancient mariners were astute observers—their trade was not only a science, it was an art. The more intuition and sense of the sea a mariner had, the better he was able to guide his ship quickly and efficiently from port to port, bringing greater profit to his employer. It is important to understand that navigation in ancient times was much less precise than it is today. A captain would use all the tools available to him—astronomical observations, soundings, estimation of the directions of winds and currents, and even the directions followed by migrating animals—to guide his ship as close as possible to its destination. Once the coastline was sighted, he would use his knowledge of the terrain to correct the vessel's heading accordingly and guide it into port.

Navigators of antiquity managed well without the advantages afforded by the compass. When the invention was finally made, its effects were more subtle than we might have expected, and yet their consequences changed the world. The compass did not enable navigation—navigation across the seas took place long before the compass was invented— but the compass made navigation much more efficient by opening the seas to winter sailing and by extending a ship's range to regions that were previously unexplored. The compass became the catalyst to the growth and expansion of worldwide trade. This amazing navigational tool brought increased wealth and national prosperity to the nations that knew how to exploit it.

THREE

━━━━►)«(●)»◄━━━━

Dante

Tᴴᴇ ʀᴏᴏᴛs ᴏꜰ ᴛʜᴇ ᴍᴀɢɴᴇᴛɪᴄ ᴄᴏᴍᴘᴀss ᴀʀᴇ ᴍʏsᴛᴇʀɪᴏᴜs. In the West, the compass was first mentioned in A.D. 1187, in the writings of an English Augustinian monk, Alexander Neckam (1157–1217). Neckam's book, *De Naturis Rerum*, contains the following description:

> *The sailors, moreover, as they sail over the sea, when in cloudy weather they can no longer profit by the light of the sun, or when the world is wrapped up in the darkness of the shades of night, and they are ignorant to what point of the compass*

*their ship's course is directed, they touch the magnet with a
needle. This then whirls round in a circle until, when its motion ceases, its point looks direct to the north.*

Neckam's text gives no hint as to where or how he might
have seen or heard about the magnetic compass. He did not
necessarily encounter the compass in Britain, for we do
know that Neckam studied for several years in Paris and that
he once accompanied the Bishop of Worcester on a trip
to Italy. Since many sources credit the Italians as the first
European navigators to use the magnetic compass, perhaps
Neckam's description is of an Italian nautical compass.

Later references to the compass in European sources are
mostly in poetry. The next mention of the compass came
between 1203 and 1208 in a long poem called *La Bible* by a
French monk, Guyot de Provins, who lived in Cluny. The
poem contains the following verses:

> Un art font qui mentir ne peut,
> Par la vertu de la magnette.
> Une pierre laide et brunette
> Où li fers volontiers se joint
> Ainsi regardent le droict point;
> Puis, qu'une aiguile l'ait touchie
> Et en un festu l'ont fichie
> En l'eaue la mettent sans plus

Shorewood Public Library

Checkout

#	Title
	Barcode
	Return Date

1 The riddle of the compass : the in...
35250004517757
2/27/2017

The Riddle of the Compass

Et le festus la tient desus.
Puis se tourne la pointe toute
Contre l'estoile, si sans doute . . .

———•❦•———

An art [the sailors have] that cannot deceive
by virtue of the magnet
an ugly brown stone
to which iron voluntarily attaches itself
touching the needle with it
they fix the needle in a straw
and float it on water
whereupon it turns infallibly
to the North Star without a doubt.

Where and how Guyot found out about the magnetic compass and its use in navigation are unknown. We know that he traveled to the Levant during the Third Crusade (1189–92), so perhaps Guyot learned about the use of the compass aboard one of the ships heading to the Holy Land.

The next European mention of the compass is by Jacques de Vitry, the crusader Bishop of Acre, who wrote in 1218 that the compass was a necessary instrument in navigation on the seas. He claimed that the lodestone was not only crucial in sailing but was also resistant to witchcraft and could be used to heal madness, as an antidote to poison, and as a cure for insomnia.

Another poem followed in the second half of the thirteenth century. This one was by the Italian poet Guido
Guinizelli of Bologna. The poem contained a description of
the magnetic needle and its property of pointing toward the
polestar. It read:

> In quelle parti sotto tramontana
> sono li monti della calamita
> che dan virtude all'aere
> di trarre il ferro; ma perchè lontana
> vale di simil pietra havere aita;
> a farla adoperare
> et dirizzare l'ago inver la stella

> *In these parts under the north wind*
> *are the mountains of the magnet,*
> *which give the air the quality*
> *of attracting iron; but why so far*
> *lasts the effect of such stone,*
> *which finds its use by making*
> *the needle point right at the star*

This was the first Italian reference to the compass. Guinizelli
was admired by Dante Alighieri, and in *Purgatorio* (canto 26,
97–99), Dante calls him: "the father / of me and of the others—those, my betters— / who ever used sweet, gracious
rhymes of love." Guinizelli was among the first poets to use

the *dolce stil nuovo,* the "sweet new manner" of poetry, as Dante referred to it. Because of Guinizelli's innovations in poetry, Dante looked up to him as to a father. Guinizelli's use of the compass—the new invention of the time—as a metaphor in poetry would be adopted by Dante himself within a few decades.

In 1269, Peter the Pilgrim of Maricourt (also known by his Latin name, Petrus Peregrinus) wrote a letter from a military camp in the province of Apulia, in southern Italy, where he was campaigning with the Duke of Anjou. He spent his time in camp writing a treatise on the magnetic compass, published as the *Epistle to Sigerius de Faucoucourt, Soldier, Concerning the Magnet.* In his *Epistle,* Peter gave a description of the dry-pivoted compass, that is, a compass in which the magnetic element is supported in the air by a pin placed under its center. He also described a floating compass, one where the magnetic element is suspended in a liquid. Peter's manuscript became the cornerstone of all subsequent work in Europe on magnetism and the compass. Three centuries later, the famous English mathematician and philosopher John Dee was to write in the margin of Peter's book that he was wrong in supposing that the magnetic needle was attracted to the polestar. It sought Earth's magnetic pole, said Dee.

Poets continued to be fascinated by the compass. The magnetic needle, mysteriously moved by an unseen force to point to the North Star, was a metaphor they could not resist. Francesco da Barberino, a Tuscan lawyer and notary who had studied in Bologna and Padua and worked for four years at the papal court in Avignon before returning to Florence, published a work of poetry in 1313 called the *Documenti d'Amore*. In rhymed Italian with a Latin translation, the poem gives rules for a good life at sea. Da Barberino gave instructions also for persons who might find themselves shipwrecked. If you find yourself in such a predicament, da Barberino says in his poem, you should build a compass. Da Barberino's text gives us the first real reference to a portable, self-contained compass, one that could actually be used anywhere at sea to help sailors find their way.

The fascination of poets with the magnetic needle is also evident in an earlier work. In 1294, the Italian poet Leonardo Dati wrote a long poem called *La Sfera* ("The Sphere"). In the third book of this poem, canto 5, Dati included the verses:

Col bussolo della stella temperata
Di Calamita verso tramontana
Veggono appunto ove la prora guata.

With a compass to the star directed
Of magnet toward the north
Come exactly where the prow points.

In 1300, the year we generally take to mark the debut of the compass with a compass card as an instrument in navigation, Dante descended into the Inferno, as he describes in his *Divine Comedy* (written a decade later). Dante lost his way in a dark wood (modern scholars interpret this event as taking place on Good Friday in the year 1300) and found his own "compass"—Virgil—who guided him down into the Inferno, through Purgatory, and finally to Paradise.

In *Paradiso,* after he reaches the fourth heaven and enters the sphere of the sun, Dante hears the song of souls. He is reminded of the songs of the sirens, who from rocky islands lured sailors away from their true course and caused them to be shipwrecked. But then he hears a voice that makes him turn, like the compass needle, toward it (canto 12, 28–30):

> Dal cor dell' una delle luci nove,
> Si mosse voce, che l'ago alla stella
> Parer mi fece in volgermi al suo dove.
>
> ————◆————
>
> *Then from the heart of one of the new lights*
> *There came a voice, and as I turned toward it,*
> *I seemed a needle turning to the polestar.*

The voice of kindness that makes Dante turn toward it is that of Saint Bonaventure, a Franciscan mystic. Dante uses the new invention of his time, the compass, as a metaphor, its needle symbolizing the attraction of the soul to righteousness and eternal love. Dante wrote these verses between

1310 and 1314. They show us just how commonly known the magnetic compass had become in Europe by the early 1300s.

The name the Italians gave to their new instrument was *bussola,* which still serves as the Italian word for *compass.* This term was used for the first time in literature in a commentary on Dante's *Divine Comedy* by Francesco da Buti, published in 1380, half a century after the *Divine Comedy* first appeared. Da Buti used the term *bussola,* indicating a boxed compass with a compass card featuring a wind rose, as compared with Dante's poetic magnetic needle pointing to the North Star.

The term *bussola* is derived from the ancient Italian word *bussolo,* which in turn is a linguistic corruption of the Medieval Latin words *buxida* and *buxus,* "a wooden container," both originating in the Classical Latin word *pyxis,* meaning "box." Francesco da Buti described the *bussola nautica,* the nautical compass, as a wooden box with a glass cover in which a round disk attached to a magnetic element rotates freely, indicating directions in degrees from 0 to 360 and including a wind rose. This wind rose featured sixteen points.

We know that the traditional nautical system of directions in the Mediterranean used twelve winds, and that this system can be traced back to classical times and remained unchanged until the Middle Ages. Da Barberino, in fact, lists within his poem twelve winds. But at the same time, nautical charts showed wind roses with sixteen directions, and all descriptions of compasses with wind roses from the late Middle Ages to modern times include sixteen directions or multiples of sixteen (thirty-two and sixty-four). When and why did the change from twelve to sixteen take place? And where did the directions used in navigation originate?

FOUR

The Etruscan Chandelier

THE FOUR BASIC DIRECTIONS USED IN NAVIGATION—
north, south, east, and west—have ancient roots. These
designations first appeared in the Bible to describe the directions from which enemies would attack Israel and the directions in which the defending armies should move.

The Land of Israel is oriented in a generally north-south
direction along the Mediterranean coast. The sea borders Israel on the west. To the east are the barren, rugged mountains of Edom. To the north lie the verdant mountains of

Lebanon, and to the south the desolate Negev desert. Early on in Israelite history, the Bible defined the four principal directions using the unique geography of the region. North in the Bible is *Tsafon*; east is *Kedem*, the direction of the red mountains of Edom; south is *Negev*, named for the desert; and west is *Yam*, meaning sea. This designation goes back at least thirty-five hundred years. King Solomon's mariners likely used these direction names in sailing in the Mediterranean and down the Red Sea three thousand years ago.

Sometime later, still long before the compass was invented, more directions were introduced in order to make the art of navigation more precise. These directions were based on winds. The wind directions led to the invention of a wind rose, which was later adopted for use in the compass. The exact way this development came about is not known.

In the center of Athens, above the famous shopping and entertainment district called the Plaka, adjacent to the Acropolis, is an archaeological site that was the *Agora,* or marketplace. At this site stands an octagonal tower, which predates the Roman-era market. It is called the Tower of the Winds. The tower displays images representing the eight winds that stood for the directions north, south, east, west, and the four directions that lay between every pair: northwest, northeast, southwest, and southeast. The tower celebrates the ancient art of navigation. It was built by the

The Tower of the Winds, Athens. *Craig and Marie Mauzy*

second-century-B.C. astronomer Andronicus of Macedonia. Each of the eight winds is represented by a male figure.

From the eight-wind system, a twelvefold direction system evolved for navigation, when four winds were added to the original eight. For mariners in the pre-compass era, directions were associated more strongly with winds than with the sun and its risings and settings. The reason for this is that in navigation the direction and force of the wind matters most to the mariner, not the position of the sun. The wind brings the weather, and the kind of weather depends on the direction from which the wind blows. "It follows," as E. G. R. Taylor notes in *The Haven-Finding Art*, "that the 'feel' of a wind gives a rough indication of direction, and it is not surprising that the names given to the predominant winds became the names for the very directions from which these winds were blowing." In the Northern Hemisphere, cold air comes from the north, and warm air from the south. Thus Boreas, the Greek name for the cold northerly wind, became the name for the direction north. Notus was the name of the warm southerly wind, so the south became Notus. Zephyr, the mild westerly wind, gave its name to the west; and Apeliotes, the dry easterly wind, lent its name to the east.

But sailors could observe and distinguish winds more precisely: there was a moist north wind and a drier one. If a north wind had a westerly element to it, it was blustery and brought rain. So now this wind was not Boreas—it was called Argestes. Similarly, other divisions were made between north and east and between south and east, and south and west. The wind-direction system could thus be seen as an eightfold system. This was the system depicted on the Tower of the Winds in Athens.

The wind rose is a diagram showing the directions of various winds. The wind rose of twelve directions was said to have been invented by Aristotle Timosthenes, a sailor-scholar who lived around 250 B.C. and was chosen by Ptolemy II, king of Egypt, to be the chief pilot of his navy. Ptolemy admired science and technology, and through his leadership, the Egyptians made great strides in many areas, including navigation. Timosthenes's twelve winds included Boreas and Notus, Zephyr and Apeliotes, and two winds between

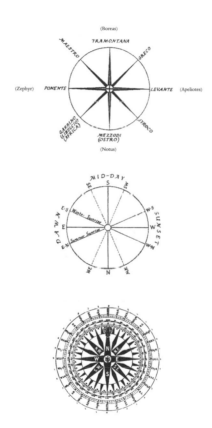

The development of the wind rose: the eight winds of
the Mediterranean (with their Latin names), the twelve winds
of classical antiquity, and a modern-day compass wind rose

each adjacent pair. These twelve directions were arranged in a wind rose.

Timosthenes wrote a book of sailing instructions for the Egyptian fleet. The book contained guidelines on navigation, and it was expanded in later centuries. Other sailing books followed, and they became indispensable tools for mariners. By the twelfth century, sailing instructions to every port throughout the entire Mediterranean were available. An example of one of the sailing instructions from Timosthenes' original sailing-instructions book is the following: "From Chios to Lesbos, 200 stadia with Notus." By comparison, here is an example of a modern sailing instruction, from *Nicholl's Concise Guide to Navigation* (1989): "Bombay to Aden: Steer SSW to 6 degrees north, then WNW to 8 degrees north, thence to Guardafui." In essence, the ancient sailing guides are strikingly similar to their modern counterparts: Both tell you which direction to follow, and for how long, in order to reach your destination in the most efficient way. The instructions in the ancient guidebooks demonstrate why a wind-based system was so useful for mariners. In the Timosthenes example, to sail directly from Chios to Lesbos, a sailing ship had to run with the southerly wind, Notus. The wind gave the ship its direction.

Wind directions may have illustrated the margins of ancient charts. Surviving medieval charts show directions in the form of a wind rose, or as puffing heads placed on the margins. Such charts, before the arrival of the magnetic compass, feature eight- or twelvefold wind-direction systems. But after the compass came into use, the wind rose was inexplicably transformed to include *sixteen* directions. Why did this happen? To answer this question, we must direct our attention to an ancient culture, one not commonly associated with navigation.

While archaeology has made great strides in modern times and has taught us much about lost civilizations, surprisingly little is known about the Etruscans. The Etruscans were a native Italic people who lived in Etruria, a hilly land of vines and olive groves, roughly corresponding to modern-day Tuscany and Umbria (although covering a somewhat wider area). The Etruscan civilization flourished from the ninth century through the first century B.C. when it was subsumed by Rome.

Until recently, the Etruscans were almost completely shrouded in mystery. We knew that they predated the Romans and that they left behind elaborate sarcophagi, indicating a preoccupation with the dead. But little else was known about this ancient civilization. Now, science has revealed more information about the Etruscans, including their language and customs. The Etruscans lived in small villages or rustic country villas for the wealthy. Later they built some of the first cities in Italy, including Perugia, Siena, Cortona, Volterra, Arezzo, and Fiesole. Etruria thus became a loose confederation of city-states linked through common language, religion, and customs.

Like the Romans, who followed them, the Etruscans were fond of feasts. These events usually started in the late afternoon and lasted well into the night. Surviving frescoes show wealthy Etruscans being served copious amounts of food by their slaves, while reclining on wooden couches. The Etruscan diet consisted of vegetables, bread, grains, cheese, fruit, and some meat.

During the eighth century B.C., the Etruscans made their first contact with the Greek and Phoenician civilizations, which were more advanced. The Etruscans adopted Greek mythology and combined it with their own cult of the dead. The Etruscans regularly made votive offerings to their gods, including body parts made of clay. They hoped the gods would bless the parts whose images were offered to them. From archaeological evidence, we can deduce that fertility was a major concern.

Orphic cults based on mysticism proliferated in the regions of Etruria. The Etruscans used augurs to prognosticate about natural events such as storms and thunderbolts; to determine the origins of these events, the Etruscans developed a direction system. This system was also used in conjuring tricks by their priests and was believed to hold magical powers.

In 1947, the Italian scholar Bacchisio Motzo made a surprising discovery about medieval navigation. He was attempting to solve a major mystery in the evolution of navigation from antiquity to the Middle Ages: What had catalyzed the sudden change from the historical eight- and twelvefold wind rose to the sixteen-point direction system featured on the magnetic compass used by Italian and other navigators from the end of the thirteenth century onward?

His research led Motzo to study ancient Etruscan divination methods. From Roman sources, he knew that Etruscan augurs divided the horizon into sixteen equally spaced points. Could this unique division system have had something to do with the wind directions and, by extension, ultimately with the compass? And why would the magnetic compass be associated with sixteen directions when traditionally navigators had always used eight or twelve directions? Motzo hypothesized that Etruscan mystical practices were linked with the use of a magnetic device, and that this device together with the sixteen points used in divination led to the construction of a magnetic compass. But the theory that sixteen points of the horizon formed the basis for Etruscan divination came from later Roman—not Etruscan—sources. What Motzo needed was an actual find, an Etruscan artifact, that would substantiate the theory that the Etruscan mystics used sixteen directions.

The Museo dell'Accademia Etrusca is located in the central square of the Tuscan town of Cortona. It is a small museum, with one large room and several smaller ones, that exhibits paintings by Renaissance artists and glassed displays of Etruscan artifacts, including many small bronze figurines of

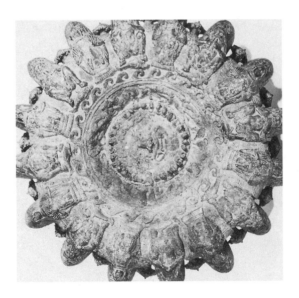

The Etruscan chandelier. *Museo dell'Accademia Etrusca, Cortona, Italy*

people and horses, as well as some jewelry. But the central piece in the museum hangs from the ceiling in the main room. It is a large exquisitely detailed Etruscan chandelier featuring sixteen figures arranged around its periphery.

The bronze chandelier, dating from between the fifth and fourth centuries B.C., was found intact in 1840 at the bottom of the hill on which Cortona is situated. Because it is unique, the artifact has received much attention from archaeologists, who continue to be mystified by the rich symbolism of its mythological figures. The chandelier is fashioned from a large, circular piece of bronze, weighing 130 pounds. In its center is a gorgon surrounded by sixteen figures: eight ithyphallic satyrs alternating with eight sirens. On the side facing the ceiling, the chandelier displays sixteen horned, bearded hollow heads of mythological creatures. The hollow heads probably held the oil that lit the chandelier.

Scholars believe that the chandelier was made by the members of an Etruscan Orphic cult that worshiped the mythological figures. While the exact meaning of the displayed creatures remains a mystery, clearly the number sixteen had relevance. Motzo and other scholars have suggested that the sixteen figures, viewed from below, represent a sixteen-point system of dividing the horizon—the same system used in modern navigation.

————

The Etruscan chandelier is not the only artifact to suggest the symbolic importance of the number sixteen. Since the chandelier's first analysis by scholars, other finds believed to have originated in Orphic cults that flourished around the Mediterranean Basin have also been studied. Interestingly, some of these artifacts of marble or pottery also exhibit a division of the circle into sixteen points.

A remarkable piece of marble called the Coppa Tarantina, a bowl that reflects the merging of Greek with Egyptian mythology, was found in southern Italy. This work of art featured sixteen deities, called *mystai,* around a circle in a design very similar to that of the Etruscan chandelier. The bowl has mysteriously disappeared from a museum in Bari, Italy.

The symbolism of the number sixteen apparently passed on to medieval times. A liturgical bowl, dated to the thirteenth century, found on Mount Athos in Greece also exhibits sixteen figures around its perimeter.

Wind directions and magnetism were linked through their relationship to mystical cults. On the island of Samothrace in the Aegean Sea, archaeologists have discovered a large marble wheel believed to have been used in divination. The wheel, called the *Arsinoeion,* is divided into sixteen sections. The find is significant since throughout the Greek world similar decorative items usually feature a division into ten or twelve elements. Scholars believe that the Arsinoeion belonged to a cult that flourished on the island from classical to Christian times. The Samothrace cult used the lodestone in divination; its members wore rings made of lodestone, which were attracted to a large lodestone key kept by the leader. Tradition links the Samothrace cult to the Argonauts and other mythological sailors. In the Samothrace cult may be found an early confluence of the three

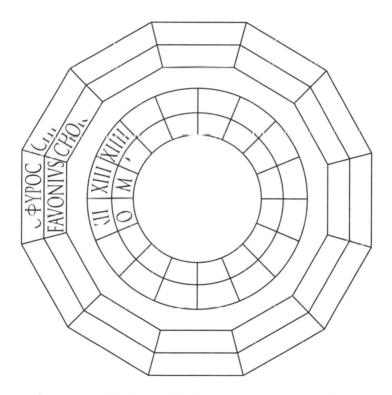

Fragment of the Prague disk showing both the twelve winds
of the Greeks and the sixteen directions of the Etruscans

elements of the compass: sailing, magnetism, and a sixteen-point system.

A direct connection between the archaic wind rose with twelve points and the mystical sixteen-point system that found its way to the modern compass card was made through an important archaeological find in southern Italy in the last century. The artifact is a marble disk now in the archaeological museum in Prague. The Prague disk includes *both* the twelve wind directions of antiquity and the sixteen directions used by the Etruscans and others around the Mediterranean. The Prague disk is therefore a kind of Rosetta stone, translating one system into the other.

The Etruscan chandelier and other artifacts provide evidence for the mystical origins of the compass. The instrument in its advanced form, used from the late thirteenth century onward, was based on a sixteen-point wind rose. These directions originated in divination cults that flourished in early times on the shores of the Mediterranean. But in southern Italy, persistent traditions link the invention of the magnetic compass with the city of Amalfi.

FIVE

⟫⟫⟪⟪

Amalfi

IN ORDER TO UNDERSTAND THE HISTORY OF AMALFI, we must first learn something about its large neighbor only twenty-five miles to the northwest: the city of Naples. Today, Naples is a metropolis—the largest city in southern Italy and the third largest city in the whole country. Naples has a magnificent bay, offering a natural shelter to ships of all sizes, including the largest American aircraft carriers. And, indeed, the well-developed port of Naples is one of the most important in the whole Mediterranean. A major harbor was built here by the ancient Greeks, and the city has been a maritime center throughout history.

In comparison, Amalfi today is a tiny village with hardly a port at all. It has no significant natural bay, only a very small breakwater behind which a few fishing boats crowd. The rickety boat that comes in twice a day from Sorrento, bringing a few intrepid souls, barely squeezes into this sheltered area. The majority of tourists come to Amalfi by bus. So the first question that comes to mind as we contemplate the history of the compass is, How is it at all possible that the boxed mariner's compass should have been invented here in Amalfi? Why not Naples? Or Venice? Or Genoa?

In the fifth century B.C., a new city—in Greek *Neapolis*, which later became *Naples*—was founded near the older Greek colony of Cumae, at the northern boundary of what we know today as the Bay of Naples. In 326 B.C., Naples and its neighbors became allies of Rome. As Rome grew to an empire, Naples and its immediate area, the region of Campania, became the favorite playground of Rome's elite: Emperors, senators, orators, and poets all sought the pleasures of Naples and its vicinity. Thanks to its wide and deep natural bay, Naples played an important role throughout history as a maritime city, and its ability to shelter large fleets gave it strategic significance from Roman times to the Cold War.

In A.D. 79, the mighty Roman fleet was anchored at Misenum, at the northern entrance to the Bay of Naples, just at the moment when Vesuvius erupted and covered the bay with ash, destroying the resort towns of Pompeii and Herculaneum. The fleet aided people fleeing the destruction by ferrying them to safer locations around the bay. This story was reported by an eyewitness, Pliny the Younger, who watched the eruption from the safety of Misenum, while his uncle, Pliny the Elder, who was the commander of the fleet, died in his efforts to save lives. The younger Pliny's account gives evidence to the great importance of the Bay of Naples as an anchorage for the Roman fleet. Today, the U.S. Sixth Fleet still patrols the entire Mediterranean from its large base in Naples.

After the collapse of Rome, Naples fell into the hands of the Goths in 543, but ten years later returned to the rule of the Eastern Roman Empire, governed from Constantinople. Naples maintained its independence throughout this period, until it was taken by Normans in 1139 and incorporated by Roger II into his Kingdom of Sicily. In 1224, Roger's grandson, Frederick II of Hohenstaufen, founded the University of Naples ("Federico II"), which is named after him. Thereafter, the arts and sciences flourished in Naples, and the city became a cultural and intellectual center of southern Italy.

During the reign of Charles of Anjou, 1266–85, Naples became the capital of the kingdom. Because of its importance as a naval and maritime center, Naples was fought over during many wars throughout the ages. In 1442, it came under the control of Alfonso I of Aragon and was the residence of Spanish viceroys until the 1700s. In 1713, Naples fell to the Hapsburgs, and in 1748, it went to the Bourbons, under whom it remained until its incorporation into the united Italy in 1860. Because of its importance as an administrative center, records of people and events in the history of southern Italy, including those of the neighboring Amalfi Coast, were kept in Naples. It would be here, at the royal archives of Naples, that the search for records relating to the identity of the elusive inventor of the compass of Amalfi would take place toward the end of the nineteenth century, spawning a controversy that continues to this day.

The collapse of Rome triggered a sequence of events. As a result of the invasions of Italy by the Germanic tribes in the centuries following the fall of Rome, the infrastructure of the empire was destroyed. Roads and bridges that for centuries connected the various parts of Italy with the rest of the empire were destroyed. The sea routes to the colonies overseas were also cut off. Only the coastal cities of Italy— Amalfi, Gaeta, Naples, and Venice—could continue their connections with Constantinople and the East. Then, in the first half of the seventh century, the Eastern Roman Empire lost Syria and Egypt. The only remaining sea commerce of the empire was between Constantinople and these cities. Finally, Naples and its neighboring Gaeta fell to the marauding tribes, leaving only Amalfi, in the south, and Venice, in the north, as maritime centers able to trade with Constantinople and the East. The time had arrived for the small Amalfi to take over control from her much larger neighbor, despite Naples's natural harbor and impressive bay, and to assume the reins of a maritime power.

According to legend, Amalfi was founded by Constantine the Great. The earliest records of its existence, however, do not appear until over a hundred years later, in the sixth century. In the Middle Ages, Amalfi was an independent city-state with a population of 50,000 and was ruled by dukes, the position becoming hereditary in later periods.

In the second half of the seventh century, Amalfi established maritime trade relations with North Africa, and in the ninth century, Amalfi reconnected with Syria and Egypt. Amalfi achieved these commercial relations ahead of Venice, and became Queen of the Mediterranean in early medieval times, assuming a title that would apply to Venice in the late Middle Ages. Even after Amalfi lost its independence to the Normans at the end of the eleventh century, it remained a powerful maritime state.

In 1071, Amalfi was conquered by Robert Guiscard, the son of a minor Norman baron, who had dominated southern Italy and went on to capture Bari and Salerno and incorporate them in his growing realm. Robert Guiscard was called "The Crafty," but he was more audacious than crafty—for he planned to take Greece and continue on to

Constantinople and crown himself the new Byzantine emperor. The reigning emperor, Alexius I, was understandably nervous when he caught drift of the Norman's ambition. The emperor enlisted the naval help of his allies, the Venetians, and in what became Venice's first major naval victory, the Venetian fleet checked Robert's progress and kept him holed up on the island of Corfu, unable to continue on to the Greek mainland. The battle demonstrated the strength of Italian maritime powers; the Italian fleets had grown and matured and would continue to do so throughout the eleventh and twelfth centuries.

In 1077, Guiscard formally incorporated Amalfi in his Norman Kingdom of southern Italy, and thereafter Amalfi rose to great wealth through maritime trade. As a sea power, Amalfi competed with Genoa, Pisa, and Venice. Naples was no longer a maritime entity of any influence. Amalfi now set the tone in all matters of navigation, including the creation of a new law of the sea. Amalfi's code of maritime law, the *Tabula de Amalpha,* became universal throughout the whole Mediterranean from the thirteenth through the sixteenth centuries. In 1206, to celebrate their new prosperity, the people of Amalfi built a magnificent cathedral in the Romanesque style and named it after Saint Andrew, whose remains were brought to the city.

Thus Amalfi managed early on to establish itself as the preeminent commercial center in the Mediterranean Basin, trading with both East and West, before other city-states had the chance to seize the opportunity to do the same. The relatively short period during which Amalfi was the maritime leader in the Mediterranean was an important one for naval history because navigation was becoming scientific and efficient. Amalfi became the source of innovations in maritime law, technology, and science. The merchants and mariners of Amalfi were adept at exploiting the temporary break in the clouds of the Dark Ages. While other city-states were in disarray and overall world trade was in decline, Amalfi's commercial ties with Arabs and Byzantines flourished because the city developed maritime prowess and was able to reach distant markets efficiently.

Amalfi's great moment of glory in world history had arrived because of a fortuitous combination of events, which this small locale was able to exploit. Thus, naval innovation and development took place here, rather than at one of the old, established naval centers of the Mediterranean. This is why Amalfi excelled in navigation before Venice, and why the compass and the navigational codes were developed here, rather than elsewhere.

As Amalfi established itself as a maritime power in the twelfth and thirteenth centuries, it built its naval strength. But Amalfi's nascent military might was used by the ruling Normans to quell rebellions in the entire region, and its navy became Amalfi's Achilles' heel and the instrument of its eventual destruction. For soon, Amalfi found itself involved in naval battles in the Bay of Naples. In one such clash, vessels from Amalfi attacked the island of Ischia in 1296 after it rebelled against the Normans. Other naval battles followed, but the people of Amalfi were often on the losing side. There is evidence that Amalfian ships burned in the Bay of Naples at the close of the thirteenth century, after a disastrous naval campaign. Amalfi was spending its scarce resources, sorely needed for trade, on naval battles against the enemies of its Norman masters.

Eventually, Amalfi was sacked by Pisan forces, lost its important trading partners in North Africa, and suffered from an outbreak of bubonic plague. During the night of November 24, 1343, an earthquake and a storm destroyed much of Amalfi and its harbor, which was never rebuilt. This is why today one doesn't see a real harbor in Amalfi, making it seem incredible that so much maritime history should have been written here. Five years after the disaster, the dreadful Black Death pandemic of 1348, which ravaged most of Europe, killing a third of its population, further decimated the population of Amalfi. The city went into decline and lost its position as a maritime power. But during its golden years, Amalfi made great strides in naval technology.

Early references link Amalfi—the preeminent maritime power of the time—with the invention of the compass. One of the first historians to make a direct reference to Amalfi as the birthplace of the magnetic compass used in navigation was the Italian humanist Antonio Beccadelli (1394–1471), who wrote, in Latin: *"Prima dedit nautis, usum magnetis Amalphis."* ("The people of Amalfi were first to use the magnet in navigation.") This verse was reproduced by the people of Amalfi on the famed *Tabula de Amalpha* establishing their "law of the sea."

Various Italian sources provide evidence that the mariners of Amalfi knew the magnetic needle as early as the first decades of the thirteenth century. Since Amalfi was a dominant naval power in the Mediterranean during the relatively short period from the twelfth century until the middle of the fourteenth century, when Venice and Genoa assumed leadership, many scholars believe that the mariners of Amalfi were the first to use the magnetic compass in the Mediterranean.

Then, between the years 1295 and 1302, a true innovation took place in Amalfi. According to medieval and modern sources, the people of Amalfi attained a "perfection" of the magnetic compass, transforming it from a needle floating in water or supported in air into the compass we know today: a round box in which a compass card with a wind rose and a division into 360 degrees rotates, attached to a magnetic element. Further evidence for the range of dates within which the perfected compass was achieved is provided by nautical charts drawn in Italy in the Middle Ages. The famous *Carta*

Pisana of 1275 does not reflect knowledge by the chart maker of the compass with degrees and a wind rose, but the ones made by the Venetian mapmakers Vesconte in 1311 and Dalorto in 1325 exhibit evidence of this new knowledge.

However, the most definitive literary reference to the invention of the mariner's compass in Amalfi was made by an important Italian historian, Flavio Biondo. Biondo was born in 1385, and lived in the city of Forlì, in the plains of northeastern Italy. In 1450, Flavio Biondo published an authoritative history of the major regions of Italy, called *Italy Illustrated in Regions*. The work was undertaken by invitation of Alfonso of Aragon, the king of Naples. In the part of his book dealing with Amalfi, Biondo wrote:

> *But it is well-known that we give glory to the people of Amalfi because the use of the magnet in navigation, which relies on the magnet's quality of orienting itself to the north, was invented in Amalfi.*

Four and a half centuries later, this reference would be at the heart of the most heated debate in Italian history of science not because of what it says but because of a controversy involving its author's first name.

SIX

—⋙—•((◉))•—⋘—

The Ghost of Flavio Gioia

I N 1901, THE PEOPLE OF AMALFI WERE BUSY PLANNING
a great celebration for the following year. Taking the latest
of the range of dates, 1295–1302, traditionally assigned to the
invention of the compass in their city, they chose 1902 as the
six hundredth anniversary of the event. The citizens elected a
planning committee, and various activities were conceived,
including placing commemorative plaques and commission-
ing a statue of the person the citizens of Amalfi believed to
have invented the glorious navigational instrument in 1302;
everyone in Amalfi knew his name: Flavio Gioia.

But beyond the name, no one in Amalfi knew anything about the man: When he was born, when he died, where he lived, what he did besides inventing the compass, whether he had a family, and—most important for the moment— what he looked like remained a mystery. This complete lack of background information did not deter the planners of the event. The sculptor would assign Flavio Gioia a face, a height, a build, and clothing (including a hood) and depict him holding a great compass in his hand and studying it with a serious expression.

Just as the planning was shifting into high gear in May 1901, an unexpected development shocked the community. A letter was published in a newspaper in Naples that called into question the reason for the whole celebration. The letter, in the May 22 issue of the *Corriere di Napoli*, was entitled "About the Anniversary of the Compass." It read:

> *In reference to an article appearing in issue No. 126 of this publication, titled "The Sixth Centennial of the Compass (1302–1902)," I allow myself to make the following points: Undoubtedly laudable is the idea of celebrating a true and ancient glory of Italy, the first introduction from China into the Mediterranean of the knowledge and practice of the precious directive property of the magnetic needle. This must have happened, with all probability, in Amalfi during the tenth century. . . .*

The Riddle of the Compass

The writer went on to question the existence of Flavio Gioia, quoting research the author had published in various scholarly journals from 1868 to 1893. The letter ended with the statement:

Respecting the desire to celebrate a centennial, albeit approximate, of the invention of the compass, the celebration should be called the ninth *centennial of the invention of the compass [alluding to the Chinese invention of the compass, which the writer believed to have taken place three hundred years before 1302].*

The letter was signed:

Florence, College of the Oaks, 19 May 1901. P. Timoteo Bertelli, Barnabist

Within a few weeks, Padre Bertelli struck again, this time with a letter to *L'Unita Cattolica* in Florence. In this letter, Bertelli exposed in detail the key results of his three decades spent researching the history of the invention of the compass. Bertelli revealed the essential weapon he had against what he called "the myth of Flavio Gioia": a theory about a missing comma.

Sixty-one years after Flavio Biondo's key reference to Amalfi as the birthplace of the compass, the next literary reference to Amalfi and the compass was made in the work of a philologist from Bologna (an Italian city in the same region as Biondo's Forlì), Giambattista Pio (1490–1565). Pio wrote a commentary on the poetry of Lucrezio Caro. In his commentary, in Latin, published in 1511 in Florence, Pio wrote:

> Amalphi in Campania veteri magnetis usus inventus a Flavio traditur, cuius adminiculo navigantes ad arcton diriguntur, quod auxilium pristis erat incognitum.

———

> *According to tradition, the use of the magnet was invented in Amalfi, in the ancient region of Campania, by Flavio, by which method the navigators direct themselves north, an advance not known to the ancients.*

The second part of the statement comes directly from Flavio Biondo. But the first part of the sentence is the most interesting:

> Amalphi in Campania veteri magnetis usus inventus a Flavio traditur ...

Padre Bertelli made the following observation: If we group together the words *inventus a Flavio* and separate them from *traditur,* then the interpretation of the entire sentence is indeed the one that took hold in Italian literature following the work of Pio and led to the name "Flavio" being attached to the invention of the compass in Amalfi. This happened, according to Bertelli, because the sentence was misread to say that according to tradition, the magnet used in navigation was invented in Amalfi by Flavio. But that was a mistake, argued Bertelli. Pio meant to say that the invention of the use of the magnet in navigation, by the people of Amalfi, was *related to us by Flavio*. Furthermore, said Bertelli, this Flavio was none other than the Flavio who made the original reference to Amalfi: Flavio Biondo. Pio's sentence, according to Bertelli, somehow lost a comma after the word *inventus,* and should correctly be read as·

Amalphi in Campania veteri magnetis usus inventus, a Flavio traditur . . .

In this form, the sentence says:

As related by Flavio, Amalfi in ancient Campania invented the use of the magnet . . .

Bertelli's reinterpretation of an ancient tradition, widely circulating in the popular press in the months preceding the planned celebration in Amalfi, caused an uproar. A number of Italian scholars were quick to issue angry rebuttals of Bertelli's claims.

The complaints against Bertelli ranged from his alleged misunderstanding of Latin syntax as used by Pio to his alleged confusion of Flavio Biondo's first and last names. The critics said that placing a comma right after the word *inventus* would have been a syntactical error that Pio, a child of the Italian humanism, would not have made. In addition, in classical Latin—the Latin used by Pio—the verb *traditur* was always used in a passive form, like *dicitur, putatur, fertur,* and was used without a controlling subject. Therefore, claimed the detractors, *traditur* could not have been used to mean that Flavio related the information.

In his letter to *L'Unita Cattolica,* Bertelli said that Flavio Biondo was already famous at the time he made his original reference to Amalfi, in 1450, and that by the time Pio wrote his *Commentary,* people in Italy referred to Biondo simply as Flavio, the way we refer to Dante Alighieri by his first name. Had Pio meant to say that Flavio Gioia of Amalfi invented the compass, Bertelli continued, he could not have referred to him simply by the first name Flavio, since there were no extant references to Flavio Gioia at the time Pio wrote or earlier. Flavio Gioia could not, therefore, have been famous enough for people to refer to him simply by his first name.

There were arguments and counterarguments, but certain facts stood out despite the challenges. One was that Pio's statement clearly looks borrowed directly from Flavio

Biondo, and that he would have had to give Flavio credit. This is exactly what he did using the verb *traditur,* maintained Bertelli. Another important fact was that references to Flavio Gioia were all much later than Pio's work and historians who followed seem to have taken their information from him, and thus from Flavio Biondo.

The references to Amalfi and the compass continued, all of them repeating the same misinterpretation of Pio (if, indeed, it was one) and asserting that Flavio was the inventor of the compass rather than the conveyor of the information about the invention, thus perpetuating the story through the centuries. The first to add to the name Flavio the last name Gioia was Scipione Mazzella, a Neapolitan historian, who in 1570 wrote a book describing the region of Naples. In the part dealing with Amalfi, Mazzella wrote:

> *In Amalfi, the year 1300 brought glory to the people. Discovered by Flavio Gioia, the magnetic compass with a chart for navigation is a necessary aid for pilots and navigators. It was an invention that was never known to the ancients.*

According to Bertelli, the fact that the name Flavio Gioia was mentioned for the first time so late in history—almost three centuries after the purported discovery in Amalfi—makes the name suspect. Bertelli maintained that this was a case of information being distorted as it progressed through time.

Flavio Biondo reported the invention of the compass in Amalfi, then Giambattista Pio quoted Biondo but left out a comma, then others misread Pio, and the error became entrenched. Finally, Mazzella added the family name Gioia.

Bertelli's opponents contended the opposite: The fact that the first reference to Flavio Gioia occurred so *early*, over three centuries before their time (1901), lent the existence of Flavio Gioia credibility. Was the glass half empty or half full? Bertelli had his theory about the missing comma, and it seemed to hold together. But his opponents had on their side the many consistent references to Flavio Gioia that appeared throughout the centuries. Or did they?

The name Flavio is of classical Roman origin. It does not appear in any registry of names in Amalfi during any period from late medieval to modern times. In 1994, Giuseppe Gargano reported that none of the sources—published or unpublished—of names used in the region of Amalfi contained the name Flavio. From the start of the fifteenth century through the end of the sixteenth, certain names of classical origin were, indeed, used in the Amalfi region. These included the Italian names Giulio Cesare (Julius Caesar), Ottavio (Octavian), Marco Antonio (Mark Anthony), Annibale (Hannibal)—but never Flavio (Flavius).

We know that the name Flavio was revived from its Roman origins and became popular again, after not being used for centuries, during the period of the Italian humanism. It was given to such people as Flavio Biondo, who lived during this era, around 1450. This fact by its own strength makes it unlikely that the name would have been given to a resident of Amalfi during the thirteenth century, a century and a half earlier.

Some sources say that the inventor of the compass in Amalfi was Flavio *Goia*, rather than Gioia. Others call him Giovanni Goia. Then there are references to Flavio or Giovanni Goya. And there are references that give his last name as Gira or Gisia or Giri. Then the first name Francesco was added to the list of names and combinations of names of the supposed inventor of the Amalfi compass. Finally, someone suggested that there were *two* inventors of the compass in Amalfi, two brothers. One was Flavio Gioia, and the other Giovanni Gioia.

In 1891, Padre Bertelli, who was studying all these names in an effort to discover the identity of the true inventor of the compass, became even more suspicious than he had been until then. He wrote a letter to Commander Bartolomeo Compasso (no relation to the compass, which in Italian is called *bussola*), the director of the State Archives in Naples. Bertelli asked Commander Compasso to search all existing documents for the period between 1268 and 1320 for any appearance of the names Flavio or Giovanni or Francesco Gioia or Goia or Giri.

Commander Compasso responded that for the reign of

Charles I of Anjou, part of the period in question, he could find only the names Roberto de Goya, nominated to be the chaplain of the town of Castel Capuano, and Bernardo Giri, a soldier living in another small town on the Amalfi Coast in 1270. The commander continued that he had made a very thorough search of all public records during the reigns of Charles II of Anjou and of Robert of Anjou, up to the year 1320, and found nothing.

"There is only one more record that may interest you," he added, "which is folio 579 in the registry of public records of King Robert. It lists the name Franciscus de Ioha. It cites the register for the year 1316, folio 203. This record, however, is missing." Could this missing record have held the key to the mystery?

Non-Italian sources also suffered from the confounding proliferation of names. Gilbert of Colchester published a book called *De Magnete* in London in 1625, in which he said that the invention of the compass took place in Amalfi by John Gioia or Goia or Goe. Other variations of the names appeared in a few sources.

Did Flavio Gioia exist? Bertelli's argument about the missing comma is compelling. When we consider the way Italian historians have for centuries copied the words of their predecessors about the invention of the compass, one can easily

The Riddle of the Compass

be convinced that a misunderstanding—due to missing a comma or to otherwise misreading a sentence—could be transmitted down the chain. Once a historian misinterpreted the information, the error would be reflected and reinforced in the writings of all future historians down the line for centuries. Pio's Latin sentence can be read two ways. Was there a tradition that Flavio invented the compass of Amalfi, or was there a tradition—reported by Flavio—that the compass was invented in Amalfi?

Then, of course, there is the problem of the names. The name Flavio Gioia only appears late in the sixteenth century, and then suddenly so many other names come out as well. Could all these names be real? Human nature makes us want to assign names to people or things. The people of Amalfi found that it was not enough to claim their glory for inventing the full magnetic compass. They wanted a name, whether or not they actually had one.

It has recently been suggested that the compass may have been perfected over a period of time: first with a floating disk, then with the compass card with a wind rose added, and then finally with the division into degrees. Other improvements in the overall design of the instrument may also have been made over a period of time. If this was the case, there may never have been a single person responsible for the invention. And if the compass was invented by one person during a single point in time, that person's name may

The statue of Flavio Gioia in the center of Amalfi. *Debra Gross Aczel*

simply have been lost. So people, centuries after the event, may well have invented a name for the inventor of the Amalfi compass, then, having forgotten or corrupted it, invented other names. Perhaps this is how we ended up with so many names: Flavio, Giovanni, Francesco; Gioia, Goia, Gisi, Ioha . . .

But it doesn't matter what name we give to the inventor of the Amalfi compass. The important fact is that some person (or series of people) in Amalfi invented the boxed compass used in navigation. We may call that person any name we like—or at least the people of Amalfi can call him any name they like. And Flavio Gioia is as good a name as any. Flavio, a rare classical name, links the inventor of the compass to Italy's Roman tradition, and Gioia in Italian means "joy."

But, as Bertelli knew and fought hard to convince everyone, the first compass was not invented in Amalfi. The inventor or inventors in Amalfi simply perfected an ancient idea by placing the compass in a box—the *bussola*—and attaching to its magnet a compass card with a wind rose and a division into 360 degrees. The first compass, a simple magnetic needle pointing north and south, was invented centuries earlier, in China.

SEVEN

Iron Fish, Lodestone Turtle

BERTELLI HAD AT HIS DISPOSAL INFORMATION ABOUT Chinese science from sources that were not accessible to the average Italian citizen--the reports of missionaries to China during the seventeenth century. A few other Western historians also possessed information about Chinese science, but for political reasons were opposed to giving the Chinese credit for the invention of the compass. The historian Joseph Needham lamented the Western bias against the Chinese, saying, "There has also been the usual tendency to presuppose that nothing of real importance could have started outside Europe." He quoted a British source of the 1800s referring to the ancient Chinese descriptions of a compass as "legends," while calling the late-twelfth-century European reference to the compass "science."

The Chinese have known the lodestone and its mysterious properties since early antiquity. And while the lodestone's ability to attract metal was well known in the Mediterranean, the Chinese also understood the lodestone's direction-seeking property.

An early story known to have been written about 806 B.C. describes the palace of Ch'in Shi Huang Ti. The palace had what must have been the first metal-detection system in the world. The entire gate of the palace was made of lodestone, and anyone who tried to enter the palace bearing concealed iron weapons would be detected because of the great magnetic pull of the gate and immediately arrested.

The Iron Age in China began around 800 B.C. During this period, bone needles were replaced by iron ones, and the Chinese first noticed that lodestone attracted the iron needles. Chinese authors have asserted that the understanding of magnetic phenomena led the Chinese to invent the magnetic compass as far back as the first century A.D., or even earlier.

Ancient Chinese literature has many references to ladles or spoons with a mysterious quality: They turned to face south. South-pointing ladles or spoons, designed to resemble the Big Dipper, were made of lodestone and indeed pointed south (and north on their opposite side), thus functioning as compasses. There is a story about the emperor Wang Mang, who was the only emperor of the Hsin dynasty (A.D. 9–23). Wang Mang's palace was overrun by the Han people in A.D. 23, and he was killed during the assault and replaced by a new Han emperor. The report of the attack follows.

> *The conflagration then reached the Chêng-Ming Hall in the lateral courts, where the Princess of the Yellow Imperial House dwelt. Wang Mang fled from there to the Hsuan Room, but the flames from the Front Hall immediately followed him. The palace maids and women wailed, saying: "What should we do?" Meanwhile, Wang Mang, dressed in deep purple and wearing a silk belt with the imperial seals on it, held in his hand the spoon-headed dagger of the Emperor Shun. An astrological official placed the diviner's board in front of him, adjusting it to correspond with the day and hour. The emperor turned his seat, following the direction of the handle of the ladle, and so sat. Then he said: "Heaven has given me virtue; how can the Han armies take it away?"*

According to Needham, all Chinese rulers from earliest times followed the practice of facing south as the imperial direction. What we read above is most probably a description of Wang Mang trying in desperation to act imperially in the face of the attack and to sit facing south. Some researchers have suggested that Wang Mang was using the Big Dipper as a pointer to that direction, but the text has a direct reference to a diviner's board with a ladle. To point to the south, the ladle would have had to be made of lodestone or magnetized iron. This agrees with other descriptions in ancient Chinese texts. The same story contains references to Wang Mang's "Ladle of Majesty," used in rituals. In all probability, we have here one of the first descriptions of a magnetic compass.

In the book *Lun Hêng,* believed to have been written in A.D. 83 by Wang Ch'ung, there is a suggestive sentence: "But when the south-controlling spoon is thrown upon the ground, it comes to rest pointing at the south." An interpretation of this text by a later writer was that the spoon was a lodestone worked by jade cutters into the particular shape resembling the Big Dipper. The spoon, or ladle, was then made to revolve on the plate of the diviner's board. The plate was marked around its edges with the names of the twenty-eight *hsiu* constellations, which the Chinese used to divide the sky. Han tombs contain fragments of the diviner boards showing the Big Dipper and constellations around the perimeter. These ancient finds make it likely that at the center of the plate was a magnetic device: a lodestone ladle pointing south.

The most convincing evidence for the development of a magnetic compass in China was discovered by the scholar Li

Shu-Hua, of Columbia University, in the 1950s. Shu-Hua found an actual ancient text, the *Wu Ching Tsung Yao,* which has been definitively dated to A.D. 1040. (The book itself was completed in 1040, but its introduction was written four years later, in 1044.) This authentic Chinese text, which survived intact, gives a clear description of an unusual device: an iron fish suspended in water. The book's author, Tsêng Kung-Liang, gives a complete and scientifically verifiable description of how the iron-fish compass was constructed and used.

A thin leaf is cast of molten iron in the shape of a fish. While still molten, the fish is magnetized by cooling it down with the tail pointing in the direction of the North Pole. The thin fish-leaf is then placed in a box containing water, where it floats on the surface. The box must be kept away from wind, so when it floats freely, the fish's head will point directly south. The final instruction is that the knowledge of the construction and use of the device must be kept strictly secret.

Melting iron and letting it solidify in the direction of the earth's magnetic field induces magnetism in the metal (this process is known as thermoremanence). Because the Chinese considered south the principal direction, the fish's head naturally was oriented toward the south. The first illustration on the next page is a reproduction of the display of the south-pointing fish from the *Wu Ching Tsung Yao.*

The second illustration (after Li Shu-Hua) is of another compass. This one is a ladle with a cup pointing south,

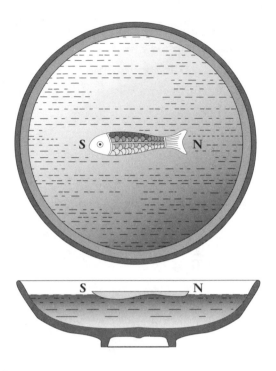

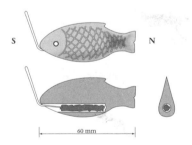

The Chinese floating fish compass

mentioned in Chinese texts and believed to have been invented during the first century A.D. This device is suspended in air rather than water.

The *Wu Ching Tsung Yao* explains in detail a sophisticated method for magnetizing the iron fish by cooling the molten

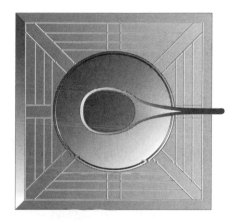

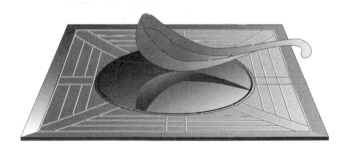

The Chinese south-pointing ladle

iron in the direction of the earth's magnetic field. A Cantonese formula for magnetizing needles consisted of heating the metal to a very high temperature over a charcoal fire for seven days and nights, adding vermilion and the blood of a rooster to the mixture. The ritual may have been developed because of the mystical qualities associated with magnetism. But the Chinese also knew how to use the lodestone to magnetize iron needles and how to fashion the lodestone as the magnetic element itself.

There are fascinating Chinese descriptions of a lodestone ladle suspended at its center on a flat mat and of a turtle suspended in the air. The book *Shih Lin Kuang Chi,* compiled between 1100 and 1250, and printed in 1325, describes a dry-pivoted wooden turtle containing lodestone with a needle—its tail—pointing south. An illustration of this delightful device (which may date from an earlier period) is shown on the facing page.

A Chinese description of the magnetic compass from a book written around 1088 shows a deep understanding of the workings of the compass. The book, called the *Mêng Ch'i Pi T'an,* was written by Shen Kua over a century before the first mention of a compass in Europe. A passage reads:

Magicians rub the point of a needle with the lodestone; then it is able to point to the south. But it always inclines slightly to the east, and does not point directly to the south. It may be

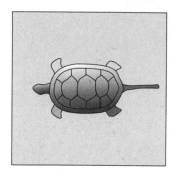

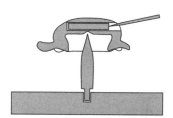

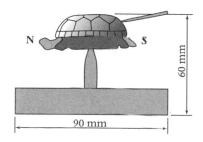

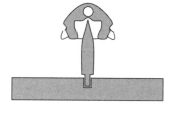

The Chinese turtle compass. *Drawing after Wang Chen-To*

made to float on the surface of water, but it is then rather unsteady. It may be balanced on the fingernail, or on the rim of a cup. . . . It is best to suspend it by a single cocoon fibre of new silk attached to the center of the needle by a piece of wax. Then, hanging in a windless place, it will always point to the south.

Just how the Chinese knew, so early in history, that the magnetic compass deviates from Earth's true geographic North and South Poles remains a mystery.

The Chinese very probably had the dry magnetic compass made of lodestone and shaped like a spoon or ladle since at least the first two decades of the first century A.D. By 1040, they certainly had a water-based magnetic compass shaped like a fish. And they had other magnetic compass designs, such as the dry-pivoted turtle and water-floated magnetic needles. But the Chinese did not use their compasses in navigation until later.

The first Chinese reference to the use of the magnetic compass in navigation appears only around 1111–17, in the book *Phing-Chou Kho T'an,* which refers to events from 1086 onward. The book contains information about sailing ships, harbors, and customs at sea. One of its passages reads:

> *The ship's pilots are acquainted with the configuration of the coasts; at night they steer by the stars, and in the daytime by the sun. In dark weather they look at the south-pointing needle.*

Chinese society was, and to a large extent still is, agrarian. China's economy is based on land and its uses, not on maritime trade. Canal and river traffic dominated Chinese navigation, and the traffic on such waterways does not require a compass. It is possible that as a land-based culture, the Chinese were not interested, in the early days after the compass was developed, in finding maritime uses for the device. They may have been more interested in the mystical powers of the magnetic needle or ladle. So before the extraordinary Chinese invention of the magnetic compass was put to use in navigation—where it was to play such a crucial role in the West—the Chinese used the device in the practice of feng shui.

Early on in their history, the Chinese had developed the "science of winds and waters" called feng shui. According to feng shui philosophy, the winds are the spirit of the earth, flowing through the veins of the land, and the waters are the waters of purification, which renew the land and its inhabitants. Feng shui is a discipline in the realm of cosmic spirituality. Its practice plays an important role in traditional Chinese culture.

Taoist philosophy, which developed early in Chinese history, led to an appreciation for every detail of topography. The shape of mountains and hills, the direction of rivers and streams, the presence of woods or grasslands, were all considered carefully. Human construction, from city walls to pagodas to residential buildings, was planned carefully to protect buildings from harmful influences and to maximize human benefit. The Chinese were concerned with the relationship between yang and yin—the opposite forces in the universe—and their influence on the terrain and its people.

Chinese paintings throughout history demonstrate the principles of feng shui in viewing and describing landscape and the layout of structures. Chinese positioning of farmhouses and city streets demonstrates adherence to the feng shui principles, and much of the beauty of Chinese art owes a debt to these principles.

Mêng T'ien, the builder of the Great Wall, once said that he couldn't build the wall without cutting through the veins of the earth. Chinese diviners were consulted in the process of making decisions about such a project, since cutting through the veins of the earth would affect people strongly.

The feng shui practitioners used the magnetic compass as a divining tool. Their approach was animistic—the compass took the form of a fish floating on water, telling them how to make decisions, or it was a turtle whose head bobbed up and down as it stabilized, pointing in a favorable direction. The diviners saw the compass's response to a force acting on it at a distance as a magical sign about the nature of the land and its water and air, as well as about what was under the earth's surface. Looking for signs on how to make their decisions, the Chinese followed the indications given them by a magnetic compass shaped in the form of an animal.

Unfortunately, much knowledge about feng shui and the compass, which could have helped us greatly in understanding Chinese culture, has been lost forever. This happened because of the interference in Chinese affairs by a foreign entity—the Church. The Jesuits, who exerted control in China in the early seventeenth century, prohibited the reading of books on many subjects, including feng shui, and Jesuit missionaries went so far as to order that books on these topics be burned. Thus many immensely valuable Chinese books fell prey to the conflict between Western ignorance and Chinese learning.

Li Ying-Shih, who was converted to Christianity in 1602, was a distinguished scholar who had amassed an impressive library with many books on divination and feng shui. He possessed important ancient manuscripts, which he had procured at great expense. These books no doubt had much information on China's civilization and culture and very likely contained details about the invention and use of the magnetic compass in divination. It took three days to burn all of Li Ying-Shih's books. Even the carved plates used in printing such books were burned by the Jesuits—to ensure that the banned books would never be printed again. Thus the European ideal of "holy ignorance" closed forever the doors to knowledge about the origins of the greatest invention China has given the world.

From what survived the burnings, we know that the twenty-four directions on the magnetic compass used in a diviner's board in China are very old and of mystical origins.

These directions date from at least as early as 120 B.C. They are associated with the tail of the Great Bear—the handle of the Big Dipper. As the time and the seasons change, the tail of the Great Bear rotates around the north celestial pole, tracing an arc in the night sky from dusk to dawn. The diviner's board, called *shih,* was segmented into twenty-eight *hsiu,* or equatorial constellations, and twenty-four directions taken by the Great Bear's tail. There was a pole at the magnetic north, as indicated by a compass needle, which was perhaps of very early origin.

Although the evidence we have shows that the Chinese used the compass in divination before they used it for navigation, we might not have the full story. It is possible that the compass was used in navigation long before the first records we have found of such use. We know that the Chinese considered their invention of the compass secret. Since their ships carried many kinds of passengers, among them foreigners and Taoist priests, who were viewed with suspicion, the Chinese may have kept their use of the compass aboard ships secret until late in the eleventh century.

At any rate, using the *Wu Ching Tsung Yao,* we can conclude definitively that the Chinese invented the magnetic compass before A.D. 1040. This date is nearly 150 years earlier than the first known reference to a magnetic compass in use in Europe. The Chinese were the first to invent the magnetic compass.

EIGHT

Venice

THE GREAT INVENTION OF THE MAGNETIC COMPASS, begun in China and perfected in Amalfi, found its first decisive role in navigation in the hands of the mariners of yet a third nation, one that in time became the greatest maritime power in history.

Venice began as a collection of small settlements on the marshy islands of a lagoon in the northern Adriatic. The early Venetians were sailors of small boats used for fishing and transport. These boats were no larger than the ubiquitous gondolas, loaded with tourists, which fill the canals today. The boats traveled the waters of the lagoon, sailed among the islands, and were used to transport goods, such as salt and fish, upriver to nearby mainland towns.

In Roman times, the marshlands and many small islands situated in the shallow northern extremity of the Adriatic Sea were called Venetia. The inhabitants of this area made their living fishing and producing salt from drying briny ponds. In ancient times the region had seven interconnected lagoons, which the Roman historian Pliny named "the seven seas." The expression "to sail the seven seas" was used to describe the great ability of the inhabitants of the islands in these lagoons as navigators. This phrase was born in Roman times—a millennium before the descendants of these skilled navigators, the Venetians, gained their supreme reputation as mariners and reached the oceans with their ships.

When the Roman Empire collapsed in the fifth century and the northern provinces of the empire fell to Germanic tribes, the coast of Venetia remained under the control of Byzantium. The sleepy islands of the lagoons and the marshy mainland with their scant population were governed for some time by officials of the Eastern Roman Empire, whose capital was Constantinople.

But Venice was soon to change from a sparsely populated collection of island villages into a maritime empire thanks to an unlikely sequence of tragic events: the successive waves of destruction of mainland Italy by marauding barbarians. The Visigoth Alaric I sacked Rome in A.D. 410, and refugees thronged the countryside. Some of them, fearing future attacks, sought refuge on the islands of the Venetian lagoons, thus enlarging the local population. According to legend, the community called Venice was formally inaugurated shortly after this first influx of refugees had settled on the islands, on Friday, March 25, 421 (although the community included several separate settlements, it did not yet include the linked islands comprising today's Venice).

While some of the refugees remained in the lagoons, many did return to the ravaged cities of Italy to rebuild their homes and resume their lives. But they were not left in peace for long. Next in the series of barbarians to sweep through Italy was none other than Attila the Hun, who stormed down into northern Italy in 452, uprooting thousands of people. Venice's population grew some more as a result of these merciless attacks, and in 466, representatives of the people of the Venetian islands met for the first time to establish a rudimentary system of self-government.

In 568, the Lombards invaded from the north, and the Roman cities of Italy were sacked and burned once again. Because of this invasion, a great flood of refugees, many of them educated and urbane, inundated the Venetian lagoons. But this time, rather than seeking a temporary refuge on the sea, most of them came to stay. Later, historians would exaggerate the aristocratic genealogy of these new Venetians, claiming they all came from patrician Roman families. However, it is true that many of the people who settled in the Venetian lagoons in 568 and after were wealthy, and many still owned land on the Italian terra firma. According to the few surviving records before the year 1000, a substantial proportion of Venetians remained landowners and were often paid by their mainland tenants in eggs, poultry, beef, or produce.

Even the famous Murano glassworks we visit today had their beginnings with Roman glass-manufacturing operations. Archaeologists have found remains of ancient glass factories on islands of the Venetian lagoon. These and other factories, as well as agricultural lands nearby on the mainland, were all in private hands. From these findings, we can conclude that Venetian capitalism was born soon after the islands were settled by Roman refugees.

The early population of Venice was mixed. The first element consisted of fishermen and descendants of fishermen who had lived in the lagoons since pre-Roman times and had, because of the nature of the area they inhabited, become competent in sailing small boats in marshes, lagoons, and rivers. Then came cultivated city people trained in the professions and possessing specialized skills, and with them, wealthy landowners. This mix of people produced a culture in which capitalist ideas and practices could thrive; Venice was eventually transformed into a society of highly capable mariners and merchants who within a few centuries would come to rule the Mediterranean world. But first, Venice would have to assert itself and challenge the dominant powers of the time. It would also need to develop the right form of government for its people.

The Italian Byzantine centers of power during this time were in Ravenna and Istria, away from the lagoon. In 697, the Venetians were organized under a separate military command of a *dux,* later known as the doge. The *dux* reported to the Byzantine powers, and Venice remained under the control of Byzantium even after the Lombards took over neighboring Ravenna in 751. The art and the institutions of Venice of the early centuries reflected its connection with and control by the Eastern Empire of Byzantium.

The region underwent drastic change beginning in 810. Charlemagne sent his son Pepin to conquer Venice. Pepin attacked Malamocco (on what today is the Lido), which was then the Venetian capital, but failed to capture the doge. The doge escaped to Rivoalto, the largest island on the main lagoon (now known as Rialto and the site of Venice's famous bridge). The Byzantines sent a fleet into the lagoon to reaffirm their control. The resulting stalemate brought about a treaty between the Byzantine Empire and the Franks. Venice, caught in the middle, was defined by the treaty as a duchy under the control of Byzantium. However, the Venetians gained more and more control over their realm as time went by, and Byzantium slowly lost its influence. The road was now open for Venice to become a republic headed by a doge and a senate.

Their near capture by Pepin alarmed the Venetians so much, however, that they began to worry that another attack someday in the future might end differently. They realized that Malamocco, situated on a sandy island facing the open Adriatic, was almost as vulnerable to attack as a mainland city. The Venetians decided, therefore, to build their main settlement on the group of islands in the center of the lagoon sheltered by Malamocco and the other *lidi* (sandbars separating the main lagoon from the open Adriatic). This decision was to change the course of history, and it demonstrated the power of maritime knowledge. With the city of Venice at its present-day location, the Venetians had an impregnable fortress. No foreign attacker could pass between the *lidi* and into the lagoon, because sailing these internal waters required a detailed knowledge of the seafloor. There were shoals and shallows that only a Venetian knew and could avoid. Once the piles and markers indicating where the deep channels passed or diverged were removed, it was virtually impossible to sail to the Rialtine Islands, the site where the city of Venice was built. Venice thus became a maritime fortress—one that would remain protected by the sea and thrive for over a thousand years.

By the ninth century, the Arabs had conquered Syria, North Africa, and Spain and established sea routes across the entire Mediterranean. A rivalry arose between the Byzantines and the Arabs, and Venice found itself in a peculiar position as the Byzantine gateway to the West. This situation became even more pronounced once the Saracens conquered Sicily and the Italian boot, leaving Venice as Europe's only door to the Levant. Venice was situated between East and West: the only connection of the Byzantine and Moslem empires in the east with the Latin-Germanic empire in the west. This unique situation offered great opportunities for trade and power, which the shrewd Venetians were not going to miss.

The East provided the Venetians with silk and spices to bring to Venice and then to transport from the lagoons up-river to the mainland, controlled by the Franks and the Holy Roman Empire. Of course, the Venetians continued to sell salt and fish there, but now their trade expanded to the luxury goods of the East.

During this time, the Venetians began to build large military vessels for defending the lagoon from naval attacks from the Adriatic Sea. These operations continued hand in hand with the expansion of maritime trade. Both merchant ships and naval vessels were equipped with newer technology as it became available. In 1081, the Venetians had a decisive military victory at the base of the Adriatic. The Venetians sent their ships to the aid of the Byzantines fighting against the Normans headed by Robert Guiscard, who by then was the ruler of Amalfi and other city-states in southern Italy. The Venetians were triumphant.

In 1082, a few months after they defeated Robert Guiscard in the Ionian Sea, the Byzantine emperor Alexius rewarded the Venetians for their help by giving them unprecedented trading privileges throughout his empire. At the same time, the emperor punished Amalfi for Guiscard's hostile intentions against him, by imposing a heavy tax on their trade with the empire. These events marked the beginning of the decline of Amalfi and the rise of Venice as a trading power in the Mediterranean.

In 1095, the pope issued an appeal to Christians to launch a crusade to retake the Holy Land from the hands of the Infidel. French and Italian nobles heeded the call, and so began the First Crusade. In 1098, a large fleet left Pisa and occupied Corfu, an island belonging to Byzantium, where it remained for the winter. The next year, another fleet left Venice and wintered in Rhodes. The Pisans joined the Venetians in Rhodes, but fighting erupted between the two fleets. The Venetians emerged victorious, and the Pisans agreed to refrain from trading in any of the ports of the Byzantine Empire.

The Venetians continued across the sea to Jaffa and arrived there before their competitors, the Pisans and the Genoese. They were there just in time to help Godfrey of Bouillon secure the ports of Jaffa and Haifa. As a result, they won more trade benefits from Godfrey and returned triumphant to Venice toward the end of 1100.

After 1100, the Venetians stopped using their fleet to defend the Byzantine Empire and instead employed it for their own use. In the following centuries, Venetian merchant ships, protected by Venetian naval vessels, dominated the eastern Mediterranean. The Crusades catalyzed the transformation of Venice into the preeminent naval power in the Mediterranean.

While the Venetians had become specialists in transporting goods along rivers from the lagoons of the Adriatic to inland Italy, cross-Mediterranean trade was carried out mainly by Greeks and Syrians and other Eastern peoples. But now Venice found itself between the empires, with a chance to transform itself from a local vassal state, involved in fishing, light industry, and short-range trading, into a maritime power.

The beginning of the twelfth century marks a turning point in Venetian history. It was about that time that the Venetians began to build larger boats. Fortuitously, the Venetians lived in one of the few areas of the Mediterranean where wood was still abundant, many other places having been deforested over the centuries. The Venetians thus had an advantage in shipbuilding. They even exported their wood to the

East, often in defiance of papal laws prohibiting the sale of wood to the Infidel.

The Venetians felt that if they were to exploit the trading opportunities of the East to the fullest extent, they had to possess the ability to construct ships more efficiently. They needed merchant ships as well as naval vessels to protect those ships against pirates and to defend the Republic against attack from the sea. Thus in 1104, the Venetians built the Arsenal (*Arsenale* in Italian), a giant shipyard that was to serve Venice well for many centuries and gave us the word *arsenal*. (The word comes from the Arabic *dār ṣinā'ah*, meaning "house of construction.") The Arsenal was built on two islands in the eastern part of Venice, and within half a century, it became a tremendous complex of dockyards, foundries, and workshops. At one time, sixteen thousand people worked there, constructing both civilian and military ships at a stunning pace. Foreign dignitaries were sometimes brought down from the Ducal Palace and taken a mile through the *Riva* to the Arsenal to admire the shipbuilding work. The King of France was once brought in the morning to see a ship's keel being laid and then again at sunset to see the same ship being launched, fully rigged, armed, and ready to sail. The Arsenal became so famous that Dante wrote about it in the *Divine Comedy*: "As in the Arsenal of the Venetians / All winter long a stew of sticky pitch / Boils up to patch their sick and tattered ships" (*Inferno*, canto 21, 7–18).

The Riddle of the Compass

As Venetian shipbuilding operations expanded, the Venetians constructed seagoing vessels for themselves and for export to other nations. Venice not only trained superb mariners who adapted themselves to long voyages across the sea but produced a merchant class as well. The merchants traveled with their goods aboard ship and were involved in making decisions about routes, timing, and directions. Venetian ships were run more democratically than other nations' ships: The person in control of the ship was called mariner, not captain, and because the merchants traveling with the mariner had a stake in what happened to the ship, they took an active role in its command. The merchants, together with the mariner and crew, would decide whether a change of course was necessary to avoid pirates, bad weather, or a dangerous rocky coast.

While their fleet was growing and their mariners began to ply their trade across the Mediterranean, the Venetians also kept their old small boats and barges for internal navigation. These boats, with their sails or oars, were used for transportation within the lagoons surrounding the islands of Venice and for connecting the various islands that did not belong to the contiguous group on which the city was built. These included the Lido (in the past there were several such sandbars, or *lidi*) and the islands of Murano (where glass factories were established away from the city for fear of fires), Torcello, Burano, and others.

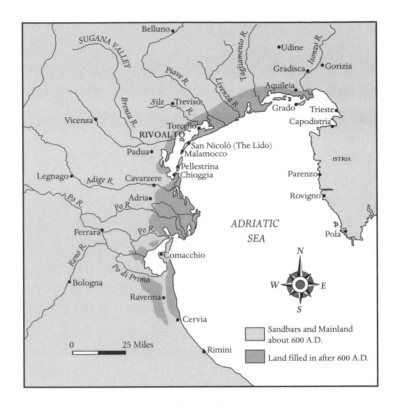

The Venetian lagoons

As they became wealthier, the Venetians intensified their participation in Crusades. At first, most crusaders traveled to the Holy Land the long way, by land, to avoid the dangers of crossing the sea. But the preferred route changed as navigation improved. The Fourth Crusade, in 1199–1204, was an important historical turning point. Instead of going to the Holy Land, the crusaders changed direction and conquered

Constantinople. A Frankish emperor, Baldwin of Flanders, was installed on the throne of Constantine, and three-eighths of the imperial city of Constantinople as well as three-eighths of the empire itself were handed over to be governed by the Venetians. The Venetians emerged from the Fourth Crusade as the unrivaled power in the Mediterranean. They were now the keepers of an empire.

In the thirteenth century, the Venetians and others sailing the Mediterranean equipped their ships with the new magnetic compass. They no longer had to waste time ashore waiting for winter to pass. Before the introduction of the compass, Venetian convoys to the Levant were timed to avoid winter: One fleet left at Easter and returned home by September. A second fleet left in August and wintered at its destination port, returning home to Venice in May. Once the compass was introduced, Venetian mariners had the luxury of navigating in a scientific way, knowing their direction precisely at all times, and using dead reckoning (using estimated speed and time traveled to extrapolate the ship's position along a known compass course) to estimate their position at sea. The innovations afforded by the compass allowed the Venetian fleets to make two round-trips a year instead of one—and neither fleet had to winter overseas. The practice brought the Venetians an immediate rise in prosperity.

The Venetians did so well at sea, both during times of

peace and war, because of their expert use of the magnetic compass. Venice changed from a small community of fishermen to the premier empire in the Mediterranean by the aid of a box containing a floating wind rose.

The change of pace in navigation took place about the same time in all seafaring nations in the Mediterranean; however, the Venetians were the foremost mariners to take advantage of the new possibilities, as they were now the rulers of the sea. Scholars have been able to pin down the date of the institution of the compass in Italy to between 1274 and 1280. Notarial records in Pisa and Genoa, as well as Venice, all indicate that in 1274, the seagoing communities still followed the old rhythm of avoiding winter sailing, while by 1290, these city-states were sending their ships to ply Mediterranean waters at all times of the year. This change was brought on by the introduction of the compass.

In the thirteenth century, the Venetian population exceeded 80,000, which made Venice one of the largest cities in medieval western Europe. Within a century, with the increase in maritime trade and the prosperity it brought, the population of the area of the "seven seas" reached 160,000, with 120,000 living in the city of Venice itself. (For comparison, outside Italy, only Paris, with 100,000 people, came close to Venice in population size.) The urban population kept growing because of constant migration from rural areas of the mainland into Venice. But the increased maritime commerce was also the cause of the great tragedy of the Middle Ages: the Black Death.

In 1347, a Venetian ship returning from the East brought with it the rats that carried the bubonic plague. Within eighteen months, three-fifths of the Venetian population perished from the disease. Other cities in Europe and elsewhere were also devastated over the following years. But Venice recovered, and was to remain a dominant maritime power for another 450 years.

The Venetians were so successful in their undertakings because they were quick to master the art and science of navigation including the use of the magnetic compass. Their maritime prowess, together with a unique and stable form of government consisting of an elected doge aided by a senate (both chosen from the nobility), ensured their continuity in a changing world. And it brought the Republic immeasurable wealth.

The advent of the magnetic compass heralded a shipping revolution in Venice. While after the year 1000 the Venetians began to build larger ships than their small lagoon boats, in the thirteenth and fourteenth centuries they built truly gigantic ships. In the Middle Ages, most ships in the Mediterranean—and elsewhere—were of less than 100 tons' displacement (roughly 80 feet long). The Venetians had a few ships as large as 200 tons, but there was nothing larger afloat. In 1260, the Venetians built a gigantic ship, the

Roccaforte. This ship had a displacement of 500 tons. For comparison, the *Mayflower* had only 180 tons, and Columbus's *Santa Maria* had only 100 tons. So the *Roccaforte* was not only the largest ship that had ever been built but the largest ship that would be seen in the Mediterranean for a very long time to come. Within a few years, the Venetians built another ship of 500 tons, and the Genoese, Venice's fiercest competitors, later built two ships of this size as well.

The reason such large ships could be built was that navigation had by then come a long way. Mainly thanks to the compass, ships no longer lost their way in fog or clouds, and precious time no longer was wasted waiting in port for the winter to pass. The use of the magnetic compass allowed navigators to practice their trade safely and efficiently. With the compass came an improved shipbuilding technology, which modernized the Venetian fleets. Thus the century following the installation of the compass aboard ships in the Mediterranean allowed the Venetians to build very large ships and to increase the number of ships they could build.

More products than ever before—salt, grain, and wine, for example—were being transported by sea. The latter two were arriving from Crete, now an important granary for the Venetians and the source of good wines. The Venetians protected their trade routes in the Aegean Sea by occupying islands, including Naxos, which became a Venetian possession. They also held territories on the Dalmatian coast and elsewhere, where their fleets needed protection from pirates. To this day, travelers find distinctively Venetian architecture at unexpected locations throughout the Mediterranean. While the rivalry with Genoa continued over the following centuries, the Venetians remained masters of the Mediterranean. Their ships transported goods from East to West and carried numerous travelers, including pilgrims to the Holy Land.

Prosperity brought with it a flourishing of art and culture. The Venetians built on their lagoon what is arguably still the most beautiful city in the world, and rich families competed with one another over who could build a more majestic palace on the Grand Canal, the city's major waterway. Among Venice's famous residents were the painters Titian, Tintoretto, and Canaletto, the composers Vivaldi, Albinoni, and Monteverdi, and the first world traveler, Marco Polo.

One of the ironies of history is that the same invention that made Venice great—the magnetic compass—was also the instrument of its eventual destruction. The Great Age of Exploration in the fifteenth century, which became a reality thanks to the use of the compass, opened new markets and new trade routes for the nations of Europe. From that point on, Venice no longer had the virtual monopoly on world

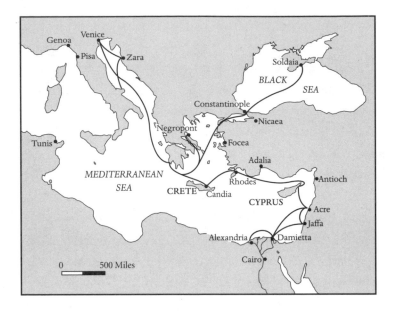

Venetian bases after the Fourth Crusade
with connecting trade routes shown

commerce it had enjoyed for centuries. From the sixteenth
through the eighteenth centuries, Venice paid less and less
attention to its fleets, both commercial and military. The
Republic's interests now centered on the Italian mainland
rather than on cross-Mediterranean commerce. But the
Venetians had always been mariners and they had neither
the know-how nor the inclination to replace their navy by an
effective land-based army. By 1797, following a lengthy period of peace and prosperity, Venice had become a city de-

voted to pleasure, with virtually no military to protect her. This was a fatal mistake. Napoleon was advancing on the Italian mainland, and the Venetians were completely unprepared to deal with the threat. They never even realized the gravity of their situation until Napoleon's forces were on the edge of the Venetian lagoon. Through his emissaries, Bonaparte was able to intimidate the doge and senate into surrendering their powers and allowing his soldiers to take the city without firing a single shot. Napoleon never set foot in Venice but continued his march through Europe. The Most Serene Republic was no more.

NINE

Marco Polo

SINCE THE DAYS OF ALEXANDER THE GREAT IN THE
fourth century B.C., there have been regular connections
between the West and India. The Indian Ocean was open to
European voyagers until the sixth century, when Cosmas Indicopleustes visited the Malabar Coast. The areas that lay
north and east of India, however, remained unknown to the
West, and Europeans are not reported to have traveled there.
This was in spite of the fact that the silk road—the caravan
trails leading from China through Central Asia to Europe—
dates from Roman times. Along this route, silk and spices

and other valuable Eastern commodities were transported across the vast continent to the markets of the Roman Empire. While the goods arrived regularly, Europeans had no direct knowledge of the mysterious country from which they came.

Theophylactus, who lived in the seventh century, was the first European writer familiar with China. He derived his knowledge of the Chinese empire from the reports of envoys who were sent to the Imperial Court at Constantinople from a Turkish kingdom in Central Asia. However, in the same century, all sea routes and land routes east to China were closed by the Moslems, then the dominant power in the Near East. Still, some traders managed to continue their operations, moving precious commodities between East and West. The Venetians, with their superior vessels and seamen, dominated much of the sea trade, but Genoese, Pisans, and sailors of other nations participated in this flourishing commerce as well. However, none of these Western traders ever ventured beyond the Mediterranean and the Black Sea. From the Mediterranean and Black Sea ports eastward, the Arabs controlled all commerce. Arab mariners sailed to China until the closing of Chinese ports in 878.

The Christian West had no direct contact with China and with the world east of Constantinople. The Crusades further separated the two worlds of East and West by causing the Moslems to unify against their Christian attackers and close the door to the East. The situation became more acute after Saladin's victory over the crusaders near the Horns of Hattin in the Galilee in 1187, after which most crusader strongholds in the Holy Land fell to the Moslems.

In 1206, the Mongol tribes of Central Asia gathered in the holy place of Karakorum and elected as *khan*, or ruler, Genghis Khan (ca. 1162–1227). Within twelve years, the Mongols under Genghis Khan had conquered the northern region of China called Cathay, and within two generations, they were masters of much of the world: Their domain spread from China to the outskirts of Western Europe. With the ascent of the Mongols, new opportunities arose. The Mongols (or Tartars, as they became known in the West because of the misspelling of the name of one of their tribes, the Tatars) were not Christian, but they were not Moslem, either. Thus many in the Christian West began to view them as possible converts to Christianity. As missionaries departed for Mongol territory, possibilities for trade opened up. The time had arrived for a Venetian merchant to seize the opportunity to trade with the vast and rich kingdom to the east.

Marco Polo was born in Venice ca. 1254 to a family that belonged to the city's merchant aristocracy. He is widely known today as the first European to have traveled to China and written about it (*The Travels,* 1298). Some who know that the magnetic compass was invented in China have naturally made the assumption that the compass was brought to Europe by Marco Polo. Unfortunately, this is not true. We know that the first European reference to the magnetic compass, in the writings of Alexander Neckam, dates to 1187— almost seventy years before Marco Polo was born.

So Polo could not have brought the first magnetic compass from China to the West. This is not to say that Marco Polo did not bring *a* magnetic compass back with him to Venice, but it would not have been the first one to arrive in Europe, and by the time of his travels, many magnetic compasses were already in use aboard European ships. In addition, if he did bring a compass back with him, Marco Polo never said anything about it in his writings. Still, the story of Marco Polo is relevant to the history of the magnetic compass. For the story of Marco Polo is a story of the relations that existed between East and West during the time navigation was expanding, and it sheds light on the state of these relations at the time the compass became important in shipping. Marco Polo's tale is also an example of how the compass might have traveled to Europe from China by the hand of some earlier, unknown traveler along the same routes, one who did not record his travels, much less publish an account of them in a famous book. Marco Polo's voyages from Venice to the East and back serve as the prototype of travels between East and West during medieval times on routes that may well have brought the compass from China to the West.

In 1255, Marco Polo's father, Nicolò, and uncle, Maffeo, two Venetian merchants who had set up business in Constantinople, sailed on the ship they owned from Constantinople to Soldaia in the Black Sea. As the owners of the ship, they had privileges in determining the route taken and in making other decisions related to navigation. From Soldaia, the two brothers continued on to the Far East in search of profitable markets. They traveled on land all the way through Central Asia to Peking, before returning to Venice in 1269. Upon their return, having made important business connections in Asia and the acquaintance of the Great Khan, they took the young Marco to accompany them on their second trip to the East.

The seventeen-year-old Marco Polo traveled with Maffeo and Nicolò to China in 1271 by land. They started from Venice, taking the usual well-traveled road to Constantinople. From there, going east, travel became much more difficult and demanding. The Polos left Constantinople, crossing the Bosporus into Asia. They trekked across the rugged mountains of Asia Minor and then made their way through the deserts of Central Persia; they braved icy storms through the Pamir passes, with surrounding mountains rising as high as 20,000 feet. Then they crossed the wide Tarim valley of western China, traveled through Turkistan, and found themselves in the southern Gobi Desert. They had to stop often to resupply and take refuge from the heat and the sandstorms, but they finally made it all the way to Peking. Their journey lasted three and a half years. The length of time and the difficulty involved with travel such as the Polos' helps to explain why China and the West remained strangers to each other for so long. And yet, the economic allure of trade in precious goods made this seemingly impossible travel a viable way of life for people like the Polos.

Soon after the Polos arrived in Peking, the young Marco became fluent in the Mongol language. His gift for languages helped the Polos negotiate deals with the people of the vast Asian empire and communicate effectively with their leader. The Polos were fortunate in making connections in the highest places: They became personal friends of the ruler Kublai Khan, whom they had brought a message from the pope. The Khan received them with great honors in his capital.

The Great Khan was taken with his new friends and in time enlisted their services in various missions. They traveled as his honored guests, armed with official golden plaques that assured them undisturbed passage throughout his vast empire, a free escort, and free room and board wherever they might find themselves within the realm stretching from the

Sea of Japan to the outskirts of Europe. The young Marco was often entrusted with special missions on behalf of the emperor. One of the Polos' most important assignments for the Great Khan was to travel west from China by sea.

A Mongol princess was to be sent to her bridegroom in Persia, and Marco Polo convinced the Khan that a sea voyage, using the navigational instruments of the time, was safer and quicker than an overland trip. The Great Khan believed that the three Polos' knowledge of navigation made them the perfect escorts for the princess. Marco Polo met Chinese pilots, discussed their art of navigation, and learned from them what they knew about navigating the Indian Ocean. He put this additional knowledge to good use throughout the trip. Marco Polo's voyage was successful. It took many months, however, and according to his account, many on board died.

Marco Polo traveled much in the East, and in total traveled more extensively than any person before him in all of recorded history. Fortuitously for posterity, Marco Polo's experiences were recorded. Years after his return to Europe, he was arrested and in prison met a writer who helped him publish an account of his travels. Thanks to this chance meeting, the Polos' travels have taken their place in the history of world exploration.

Much of the time the Polos traveled by sea; their return trip to Europe was almost entirely a sea voyage. During his years in the East, Marco Polo became an expert mariner. He understood well the Chinese methods of navigation and perfected his knowledge through close contact with navigators and seafarers of the East. And yet, tantalizingly, there is no mention of the magnetic compass anywhere in Marco Polo's writings. Is it possible that during this late period, the Chinese compass was still used mostly in feng shui?

According to Marco Polo, Chinese navigators used the North Star starting at Cape Comorin (latitude 8 degrees north) and points northward. At the cape, the North Star was barely visible over the horizon. English navigators, upon reaching Cape Comorin, would say that they "lost the pole." At this latitude, therefore, the magnetic compass would have become important for navigation of vessels sailing farther south. Marco Polo recorded two more angular heights for the North Star, one at Malabar and one at Guzerat. Polo's account tells us about the use of the polestar not only for finding directions but also for estimating a ship's latitude.

Some experts believe that the magnetic compass may have been developed in Europe independently of its invention in China. Marco Polo's story can provide us with a perspective on such theories, as it illustrates several facts about the connections between East and West. The sea routes from China to the West were difficult for passage, and were closed over

The travels of Marco Polo

many centuries. The land routes through central Asia, however, remained a viable option throughout history, and were
the preferred route to Europe even when the sea routes were
open. We know that caravans from China brought goods
to the Roman Empire, and that these same land routes, while
often disrupted by highwaymen, uncooperative Moslem

government officials, and other political factors, were still technically passable throughout the Middle Ages.

We know that the Chinese may have possessed a magnetic compass as early as the first century, or even earlier. In those days of the Roman Empire, the trade routes were open and regularly brought silk and other goods to Europe. We also know that divination cults thrived both in the Mediterranean and in China during this period. A cult on the island of Samothrace—geographically close to Constantinople and hence to the destination for all goods from China—was especially active and indeed used magnetic elements in its rituals. It is therefore plausible that over centuries of trade, a Chinese magnetic compass used in divination arrived in Europe and found its way to one of the Mediterranean cults.

Marco Polo's travels prove the feasibility of transport between China and the West. His journeys underscore the likelihood that sometime between the Roman era and his own period a compass would have arrived in Europe among the many goods that traveled the routes he and his father and uncle took in the late Middle Ages.

But why should we believe that the compass did come to the West from China? The best reason is given by Li Shu-Hua:

> One can remark that, in the work of Bailak [the first Arab writer to mention the compass], the south or "noon" is men-

tioned before the north in practical use, both in the sea of Syria and in the Indian Ocean. This detail leaves hardly any doubt that the Arabs had adopted the practical method used in China. The aquatic compass or needle floating on water, which Bailak encountered in 1242 in the Syrian Sea, and which served the French navigators during the reign of Saint Louis (1226–1270), is exactly of the same type described in the [Pên Tshou Yen I] by [Khou Tsung-Shih] around 1116. Likewise, the iron fish, used in the Indian Ocean, of which Bailak speaks in 1282, is exactly of the same type that was described in 1040 in the [Wu Ching Tsung Yao].

A bare minimum of 147 years (between 1040, the confirmed compass date for China, and 1187, the time of Neckam's work) seems quite a reasonable length of time for the invention to have traveled from China to the West. Caravans bearing silk and spices traveling the routes taken by Marco Polo could easily have carried a compass sometime before the thirteenth century. The assertion, of course, is not a certainty. It is still possible that the European compass was invented independently of the Chinese compass. Still, the Chinese were the first to invent the magnetic compass, even if they did not use it in navigation until the device had become standard equipment aboard European ships late in the thirteenth century. This may explain why Marco Polo, who wrote so much about all aspects of Chinese navigation, never mentioned the magnetic compass.

Marco Polo died in Venice in 1324. For many years, people believed that his stories were mixed with fantasy. In fact, his fellow Venetians affectionately named him the *Milion*, because, they said, he always talked in millions. Marco Polo's house—bought with the profits from his voyages—can still be visited in Venice today. The Venetians named it, appropriately, the *Corte Seconda del Milion*.

There is a legend that when the Polos returned home after a quarter century of travel, they simply knocked on the door. The servants looked through the peephole and saw three disheveled, shabbily dressed men outside. They asked who they were, and were answered: "The masters." The bewildered servants let them in, whereupon the three Polos ripped open the linings of their clothes and released a flood of emeralds, rubies, and diamonds.

Despite the legends surrounding him, Marco Polo's writings have emerged as essentially factual. Research conducted in Chinese archives and elsewhere over the centuries has confirmed many of the stories in Marco Polo's book. Today we know that his descriptions of China and the Mongol Empire he visited, as well as of the customs and lifestyles of the East at the time of his travels, are exceptionally accurate.

TEN

Charting the Mediterranean

HAND IN HAND WITH THE COMPASS, CHARTS OF THE sea and pilot books were also produced in Europe in the late Middle Ages. The wind rose could be neatly drawn on the margin of a chart, making a one-to-one correspondence between directions at sea and directions on the chart. Pilot books gave the navigator directions on how to use the chart and compass to go from one port to another in the safest and most efficient way.

Pilot books have a long history in the Mediterranean,

going back to ancient times. These directions were concise. A third-century pilot book, the *Stadiasmus of the Great Sea,* includes the following guidance for navigation to Crete:

> *From Casus to Samnonium 300 stadia. This is a promontory which extends far to the north. There is a temple to Athena and an anchorage with water.*

The invention of two principal navigational tools took place in Italy between 1250 and 1265. Both implements are believed to have been created by the same anonymous originator— and both were called *Compasso* but were not compasses. (Remember that *compass* in Italian is *bussola*.) The first *Compasso* is a *carta*—a nautical chart of the Mediterranean, while the second *Compasso* is a *portolano*—a book of sailing instructions for the Mediterranean Sea. The two navigational aids, used by mariners together with the new full magnetic compass with wind rose, brought about a revolution in world navigation. Seven centuries later, this revolution would result in the emergence of an integrated world economy.

The nautical chart was the major contribution in the Middle Ages to the field of geography. The ancients had maps of the land and sea, but these drawings were not accurate and not to scale; therefore they were not true navigational charts. With the advent of the magnetic compass that rotated and featured actual directions on a wind rose, more accurate charts were drawn in the thirteenth and fourteenth centuries, which could be used together with the new compass. The charts showed the compass directions in their margins, allowing integration of the two navigational aids. With them came the portolano, a book of sailing directions between all Mediterranean ports.

Both the portolano and the chart of the *Compasso* were improved and expanded from their initial anonymous-author form. In 1490, Alvise Ca'da Mosto of Venice published what became a widely used portolano. This was an expanded book of sailing instruction for the Mediterranean used by Italian navigators.

Inaccurate early charts of the Mediterranean were replaced by an accurate and more correctly scaled chart, which was found in Pisa (and was likely made there), called the *Carta Pisana*. The *Carta Pisana*, the oldest surviving maritime chart, dates to ca. 1275. It demonstrates an extensive knowledge of the Mediterranean Sea as well as a surprising understanding of the mathematics involved in chart making. The chart on the following page uses sixteen wind directions.

Drawn to scale with extraordinary precision, the *Carta Pisana* features a legend giving the scale and a wind rose. The linear scale on the *Carta Pisana* shows a length of 200 miles divided into four segments of 50 miles each, two of which are further subdivided into increments of ten and

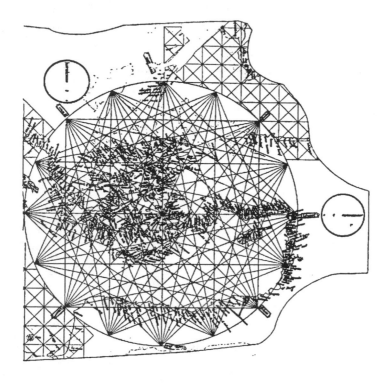

The *Carta Pisana*

five. The scale is drawn both vertically and horizontally, as on charts made today.

Navigating with the chart entailed sophisticated mathematical skills. The chart and portolano provided directions from port to port throughout the Mediterranean, and a navigator had to know how to use these directions together with his compass. The design of the compass points as they relate to the chart is interesting. From the center of the circle on the map, sixteen rays representing the sixteen half-winds were drawn. Different winds were drawn in different colors to make them easily identifiable. From each of the points at which the rays touched the circumference, lines filling one quarter of a circle were drawn. These ran to the opposite side of the map. Cities and ports were distinguished on the chart by flags painted beside them, with the emblem or coat of arms of the ruler.

Other precise charts of the Mediterranean followed. The most famous mapmaker of the fourteenth century was Petrus Vesconte, whose charts of the Adriatic Sea and other parts of the Mediterranean exhibited fine detail, such as had never been seen before. Another feature of Vesconte's charts was an accurately oriented compass rose, which allowed for the efficient use of the compass with the chart. Pictured on the following page is an illustration of a chart and compass rose from the atlas of Petrus Vesconte, 1318.

The charts that survive from the thirteenth and fourteenth centuries show compass roses similar to the one shown from Vesconte's atlas. From these drawings, scholars have deduced that the compasses with wind roses that evolved in Italy and the rest of Europe during those early years were all based on sixteen wind directions. Later European compasses used simple multiples of sixteen elements: thirty-two and sixty-four.

A chart with sixteen points of the compass from the atlas
of Petrus Vesconte, 1318. *The British Library*

Cartography flourished in Venice in the fourteenth century. Drawing a chart was truly an art, as well as a science.
These charts were drawn by hand and incorporated every
detail that would be of interest to the navigator. The charts
were then copied by hand and distributed to navigators. A
good chart was among a mariner's most precious posses-
sions. Vesconte had two gifted successors, the two Venetian

brothers Marco and Francesco Pizzigani, who also created very accurate charts of the Mediterranean.

From the records that survive of lost ships and their inventories in the late thirteenth and fourteenth centuries, we find that these ships carried charts, sometimes called *mappamundi*—map of the world—as well as a compass and spare lodestones. Charts, sailing instruction books, and a good compass had by then become standard navigational equipment aboard ships.

The Arab writer Ibn-Khaldūn noted in 1377 that all the countries surrounding the Mediterranean Sea were drawn on maps called *Compasso*. Each chart was a sheet of parchment with a compass rose on it and a drawing of the coasts of the Mediterranean. The Arabic word for the chart was *Kunbâs,* clearly derived from the Italian.

In the Mediterranean Sea, positions of ships could tolerate the uncertainties introduced by the method of dead reckoning. Here, the course of a vessel was more important to identify than its exact position, and the compass was crucial for such determinations. The immediate consequence of the invention of the compass in the Mediterranean was that navigation could continue throughout the year.

When the use of the European magnetic compass with wind rose became common, at the end of the thirteenth century, the age-old tradition of pulling ships ashore for the winter ended, and Mediterranean city-states, led by Venice, adopted the new practice of sailing all year. Genoa changed its laws, now making it compulsory for its own fleets to make two voyages a year, with one voyage beginning in February—in the dead of winter—and with no layovers overseas. Notarial records of Pisa indicate that by the 1280s ships were leaving for their cross-Mediterranean voyages at all times of the year.

The prevailing winds in the Mediterranean made the advantage of winter sailing afforded by the compass even more decisive, because ships used sails and depended on the winds. To reach Italian ports between May and October, ships returning from Egypt had to take a roundabout route through Cyprus or Rhodes because of the prevailing northerly or northwesterly winds. This was the same route taken in antiquity by the Roman grain ship, the *Isis*. But with the advent of the compass, the situation changed drastically. Because the winter winds were much more favorable to navigation on the return from Egypt to Italy, ships could follow a more direct, faster route. In October and November, prevailing winds off Egypt were easterly, allowing the Venetian, Pisan, and Genoese fleets to make the return trip more efficiently.

A surviving log of a sixteenth century Venetian ship indicates that it returned from Alexandria by a direct route south and west of Crete. It left on October 21 and reached Corfu, on its way back to Venice, on November 7, 1561. Such a voyage would not have been permitted before the introduction of the magnetic compass.

The compass, together with detailed charts drawn to scale, and a good pilot book listing all the ports and directions between them, changed forever the nature of sailing in the Mediterranean. These advances brought about unprecedented progress, characterized by an expansion of trade for all the nations on the shores of the sea and those farther inland that maintained Mediterranean fleets.

ELEVEN

A Nautical Revolution

THE MAGNETIC COMPASS, TOGETHER WITH NEW CHARTS and pilot books, also opened the way to the exploration of the oceans beyond the Mediterranean world. Mariners equipped with good compasses during the Great Age of Exploration brought about a true revolution in world commerce. This revolution changed the world.

Within a century of its debut in the Mediterranean, the compass made its way to northern Europe. As noted earlier,

Norse navigators had been able to sail to Iceland as early as the tenth century, without a magnetic compass. But navigation in the northern seas improved and expanded once the magnetic compass became standard equipment on ships, in the fourteenth century.

In the Baltic and the North Sea, the compass was also less important than it was in the Mediterranean. On the fifteenth-century chart made by Fra Mauro, there is a legend by Germany: "In this sea they do not navigate by compass and chart but by soundings." This doesn't mean that the compass was not known to the northern Europeans, but that the variations in sea depth were sufficient for navigation. Because of the depth of the Mediterranean and Atlantic, the compass was more useful since soundings would have been impossible in the open sea. In other areas, the sounding line was still immensely important as late as 1449, as demonstrated by the case of a Danzig ship bound for Lisbon that year. The ship was impounded in Plymouth, England, and to prevent its possible escape, the ship's captain was made to surrender his sounding line.

For sailing routes in northern Europe that left the continental shelf and entered deep waters, the compass was an invaluable aid in navigation. Such routes included crossings from Spain to England and the Channel area. A book of sailing instructions for these waters, compiled in the fifteenth century, demonstrates how soundings were combined with compass readings in navigating these seas. Condensed and put into modern English, the instruction book reads:

> *When you come out of Spain, when you are at Cape Finis-terre, set your course north-northeast. When you reckon you are two-thirds of the way across to England, if bound for the Severn go north by east until you come into soundings. If you then find 100 fathoms depth or 90, go north until you sound again and at 72 fathoms find fair gray sand. That is the bridge that lies between Cape Clear and the Isles of Scilly. Then go north until you come into soundings of ooze, then set your course east-northeast.*

A similar sea route on which soundings could be used only part of the way and compass navigation had to be used as well is the England-Iceland line. For ships traveling from England to Iceland during the fifteenth century, there is a mention of a "needle and stone" used in navigation. Similar sailing instructions used soundings while close to shore, and compass directions when the ship was far out at sea. During the same century, fishing boats in the Atlantic were using the compass.

The Portuguese, led by Henry the Navigator (1394–1460), became the originators of the next advances in navigation. They undertook extensive explorations of the coast of West

Africa. In the fifteenth century, they colonized the Azores and explored the coasts of the Atlantic Ocean. They made an attempt to find a way around the Islamic Empire at the eastern edge of the Mediterranean to the rich trading destinations of the Orient. Henry the Navigator was said to have established a school of navigation at Sagres, where the best navigators and astronomers were trained. Here, using astronomical observations, the Portuguese developed advanced navigational methods that would become the precursors of all modern navigation before the advent of satellites and the Global Positioning System (GPS).

The Spanish and the Portuguese made important advances in navigation in the centuries that followed the popularization of the use of the magnetic compass. Using celestial observations, they also used the compass together with the astrolabe (a precursor of the sextant) and other navigational tools to estimate positions. Of course, since good chronometers were not available, the determination of longitude would have to wait. Still, the use of the compass and a rudimentary understanding of celestial navigation allowed the Spanish and Portuguese to explore lands far beyond Europe. Their navigation improved so much with the advent of the compass that Pope Alexander VI had to be called upon to choose a meridian as a dividing line between Spanish and Portuguese territories.

The new Portuguese methods of navigation inspired Christopher Columbus. Columbus tried to use these advanced methods of navigation, but he had not mastered this art. Throughout his voyages of discovery to the New World, Columbus relied exclusively on his magnetic compass for navigation. His method of dead reckoning to estimate his position using the compass was done by multiplying estimates of the speed and time traveled, and extending the result along the compass course from the previous known position, correcting the answer by factoring in the estimated leeward drift of his ship. The calculation gave him an approximation of his position at any moment in time.

Columbus was so gifted in the use of the simple navigational method of dead reckoning that he could find his way across thousands of miles of open sea to locations where he had arrived on earlier voyages by simply following the course he had taken previously, without the aid of anything as complicated as celestial navigation. Later voyages made by other explorers across the Atlantic Ocean, as well as the Pacific and other seas, used the more advanced methods of navigation; but none of these methods would have worked without a crucial element—the compass.

In the Indian Ocean, the situation was different. Here, the invention of the compass made a less significant contribution. In the Indian Ocean, the regularity of the monsoon winds gave mariners a good sense of direction. Even under cloudy skies, navigators could rely on the monsoon and its constant direction and thus did not need a compass. In addition, skies over the Indian Ocean are clear most of the time, so the difference between summer and winter navigation is not as significant as it is in the Mediterranean. Mariners sailing between India and Arabia had little difficulty keeping their direction and relied on the compass less heavily than did those in the Mediterranean.

For voyages from Persia to Zanzibar, which entailed a north-south route, Arab navigators learned to estimate their positions and directions using the stars, so the compass was less important here as well. A century after the compass had become standard equipment on ships in the Mediterranean, European explorers reported that fleets in the Indian Ocean did not use it. The compass was little needed to determine directions there, and it wasn't used for dead reckoning since ships' positions were estimated by changes in latitude computed from the altitudes of stars.

During the time of Henry the Navigator, the Portuguese made a concerted effort to discover new trade routes and new connections between oceans, with the aim of expanding their hegemony. They had made many inroads into the unexplored oceans, but one of Portugal's greatest triumphs took place toward the end of the fifteenth century. It was the inauguration of a sea route to India that followed south along the African coast and rounded the Cape of Good

Hope, from there continuing northeast into the Indian Ocean. This great achievement was the work of Vasco da Gama (1460–1524), who sailed his fleet from Portugal in 1497. Da Gama set sail directly to the southern tip of Africa. His voyage required tremendous planning and incorporated the results of years of experience gained by previous navigators sailing down the African coast. Charts of these waters had been drawn, and compass bearings recorded. Due to da Gama's proficiency as a navigator, and to the technical perfection of his instruments, the voyage was a success.

The amount of preparation for da Gama's voyage was unprecedented. Two of the ships, the *São Gabriel* and the *São Rafael*, had been built specifically for this voyage. The ships were especially strong and were fitted with small cannons. This was unusual for a voyage of discovery; most ships used for exploration were unarmed. But because these ships were to travel distances far greater than the ones traveled before and would be away from friendly shores for long periods of time, it was deemed prudent to equip them with arms for protection.

Da Gama had two other smaller ships in his fleet, a caravel and a store ship used for carrying supplies. The crews were selected very carefully. Da Gama himself was chosen as captain general by order of the king, who passed over others possibly more deserving and more experienced. Da Gama had not previously commanded a sailing fleet, and he was officially described as a gentleman. But he showed promise as a navigator because he had a good knowledge of mathematics and the new scientific methods for determining a ship's position based on observations of celestial bodies, which were only beginning to be developed during the Great Age of Exploration.

The voyage began with what for the Portuguese had become a routine run down the coast of Morocco and on to the Cape Verde Islands. Da Gama's ships became separated by fog and sustained some damage, but all of them were commanded by captains who were good navigators and could run a compass course directly to the Cape Verde Islands despite the fog. There, da Gama had the damage to the ships' sails repaired, and the fleet continued south, using a straight-line course steered by compass for a hundred miles, until they reached a point off the Sierra Leone coast.

At the latitude of Sierra Leone, da Gama made a bold and unprecedented move. All previous voyages of the Age of Exploration had continued south by east, following the coast of Africa. This would have been the logical choice, and the one that would have kept his ships not too far from land throughout their route down the Gabon-Congo-Angola coastline toward Guinea. But da Gama, who had no experience commanding such ships, was not bound by tradition. Trusting his instincts and his ability to navigate by compass and stellar observations, he decided to do the unexpected.

Da Gama chose to turn west-southwest and sail straight into the open Atlantic.

His gamble paid off. Prevailing winds would have made it difficult and tedious to try to sail southeast. The course out to sea that da Gama chose allowed him to take advantage of favorable winds and to sail efficiently south toward the Horn of Africa. In so doing, da Gama set the trend for navigation to India for the next three hundred years. Reaching the equator at about 24 degrees west, da Gama encountered the southern trade winds and changed course to southwest by south, allowing him to take further advantage of the winds and make good time all the way south. From the Cape Verde Islands to the southern tip of Africa, da Gama's fleet was out of sight of land for over three months. This was by far the longest open-ocean passage European ships had made up to that time. Such a long crossing, with several changes of direction in the open sea, would not have been possible without expert use of the magnetic compass.

The southern tip of Africa was not a new place for the Portuguese. An earlier explorer, Bartholomeu Dias, left Portugal in 1487 and sailed down the African coast. He ventured farther along the coast than any previous Portuguese or other European navigator. Dias had made it all the way to the Cape of Good Hope.

Having passed the Cape, and after stopping at locations in its vicinity to resupply his ships, da Gama entered a territory that had never known Europeans. While in the area of the southern tip of Africa, da Gama found himself off the Pondoland coast at Christmastime. He named the land he discovered Natal. Da Gama put out to sea for a week, but was driven back by winds. After spending a month ashore, the fleet passed Sofala and, with a favorable wind, in six days found itself close to the town of Mozambique. Throughout these areas, da Gama and his crews traded for food and water with some garments and utensils they had brought with them. But as they continued up the east African coast, there was little interest in the goods they had on board. People here had been trading with India and China and were used to fine silk, cotton textiles, porcelain, and similar goods. They had no taste for the crudely made European products these sailors brought with them for the purpose of procuring food.

At Malindi, da Gama picked up an experienced Arab pilot, Aḥmād Ibn-Mādjid, who was the author of many sailing-direction books and guides to navigation and was by then, in 1498, an old man. Ibn-Mādjid helped da Gama cross the Indian Ocean in twenty-seven days, and with relative ease. A pamphlet, bearing the date July 1499, was issued by King Manoel of Portugal to commemorate da Gama's great success in navigating a Portuguese fleet all the way from Portugal to India and back for the first time in history. A new and important all-maritime trade route between Europe and India had just been inaugurated.

Da Gama returned to Europe at an auspicious moment. Because of a series of events, mostly political in nature, the cost of spices in Europe had climbed dramatically at the end of the fifteenth century. Venetians, Genoese, French, and other Europeans had come to depend on a supply of spices brought over land routes from the Orient. These routes had now been obstructed or closed, and spices and other commodities from the East were at a premium. Within a short time, Europe grasped the significance of what da Gama had achieved and the great possibilities for world trade that his accomplishment made possible.

In 1499, a hundredweight of pepper at the Rialto market in Venice cost eighty ducats. In India, the same amount cost three ducats. This tremendous price discrepancy, and similar ones for other goods, made the potential for direct sea-contact between Europe and the Orient exceptionally desirable. Within a year of da Gama's return to Europe, a fleet of thirteen ships, owned in part by the Portuguese Crown and in part by a consortium of Portuguese and Florentine merchants, left Portugal for India. This fleet was commanded by Pedro Alvarez Cabral. Cabral lost six ships to storms in the Atlantic, but made it to India. His voyage demonstrated the profitability of trade in spices by the sea route down the African coast and on to India as pioneered by da Gama.

In the following years, Portuguese and other Europeans sailed farther east. Diego Lopez de Sequiera landed in

Malacca in 1508. In 1505, the Italian mariner Lodovico di Varthema traced the steps of Marco Polo and passed through the Strait of Malacca to Sumatra and the Spice Islands. Europeans established bases in Malacca, Java, and other locations, where they could resupply their ships and arrange for goods to be stored or shipped back to Europe. The Portuguese even set up a base at Macao, on the Chinese coast. Only recently did this Portuguese colony return to Chinese control.

At the same time that da Gama made his voyage to India, Amerigo Vespucci, a Florentine businessman with a passion for navigation, made two voyages to the Americas, following in the footsteps of Columbus. One of his trips was in the service of the Portuguese, and the other for the crown of Spain. Upon his return from Central America and the Caribbean in 1500, he became pilot general of Spain. Both Spain and Portugal expanded their explorations of sea routes to the Americas in search of gold, pearls, and other valuables at the same time that navigation to India was establishing European trade routes to Asia.

The most ambitious voyage of the Age of Exploration began in 1519 when Ferdinand Magellan set sail from Cádiz with a fleet of five ships. Magellan was Portuguese but sailed under the flag of Spain. He chose mostly Portuguese officers, and a polyglot of sailors. Magellan headed south, according to plan, and reached the Cape Verde Islands. From there, the fleet crossed the Atlantic and arrived in Brazil, where Magellan relieved one of his captains of his duties because of a dispute about the route Magellan had chosen. The crews examined the Plata estuary and continued south to Patagonia, where they spent the winter. Here, a part of his crew mutinied, led by the Spanish officers. Magellan regained control of his fleet and executed the mutineers.

In Patagonia, Magellan entered unknown waters and had to rely on his navigational skills, his compass, and the stars. Surviving records of the voyage indicate that Magellan computed his latitude with excellent accuracy and that his compass directions were equally impressive. Magellan even made surprisingly good estimates of his longitude. His crews kept meticulous records of their observations of the night sky and noticed two nebulous patches during clear nights, never before seen by Europeans. We now call them the Small and the Large Magellanic Clouds, and we know that they are satellite galaxies of the Milky Way.

The Strait of Magellan, through which the fleet crossed from the South Atlantic into the Pacific Ocean, is arguably the most treacherous sea passage in the world. It is viciously stormy, with strong, unpredictable currents. The eastern entrance to the strait is deceptively peaceful, with low grasslands on both sides. But as a ship progresses through the 310 nautical miles from the eastern entrance to the western opening of the strait, the conditions change radically. The western part of the strait is a narrow fjord—a deep opening between high, ice-capped mountains. Here, the forces of nature converge on a vessel with unequaled ferocity. The prevailing westerly wind, rounding the mountains of the western edge of the continent of South America, gusts and turns the waterway, at places only two miles wide, into a stormy sea. Since there are no shelters along the way, the mariner has no place to hide. Added to the fierce winds are violent currents. The waters flowing along the coasts of South America converge here at the tip and create a cauldron. Navigating the Strait of Magellan is perhaps the most harrowing experience a mariner might endure, and the fact that Magellan's sailing ships could make it through the passage is, itself, an outstanding achievement.

Magellan's fleet had little opportunity to resupply before entering the strait. Water they received from native people on the east coast of the continent was of poor quality because of its high salt content, and the food they were able to procure was not plentiful and consisted of fish and seabirds. As the fleet entered the strait, a group of officers had second thoughts. One of the ships, the *San Antonio,* mutinied and returned to Spain with its captain and crew. But Magellan persisted with his remaining ships, traversing the entire length of the strait in thirty-eight days. Future expeditions would sometimes take several months to pass from the Atlantic to the Pacific side of the strait, although within the same century, Sir Francis Drake would set a record for sailing ships, going through the Strait of Magellan in a mere sixteen days.

Magellan celebrated his fleet's achievement by firing the ships' cannon in a salute to the South Sea. The fleet then faced the emptiness of the Pacific Ocean—an area equal to the total land mass of Earth. From then on, Magellan and his crews did not see land, with the exception of two uninhabited islands, for almost four months. They made landfall in Guam, a battered fleet of starving, half-dead men who had survived the longest sea passage in history by eating rats and water-softened pieces of wood.

Throughout the voyage, Magellan depended almost exclusively on his magnetic compass. Emerging from the Strait of Magellan, he set course north by northwest up the Chilean coast. According to a surviving logbook, at about 20 degrees south, Magellan changed course to northwest, taking advantage of the southeast trade winds. At about 15 degrees south, Magellan again changed his compass heading to west. Then another change of direction was made to northwest, making the fleet cross the equator at about 154 degrees west. At about 12 degrees north, the course was set to west and remained so until the fleet's arrival in Guam. Magellan's adept navigation by compass and celestial observations proves that a good navigator can traverse very large distances in open ocean and obtain good estimates of position without knowing the exact longitude.

The fleet continued to the Marianas and from there to the Philippines. In the Philippines, Magellan made the mistake of getting involved in local politics, favoring one ruler over another. A skirmish resulted, and Magellan was killed on the beach of one of the Philippine islands. The remnants of his crews continued west and arrived at Tidore in the Moluccas in November 1521. A Basque captain, Sebastián del Cano, assumed command of Magellan's last remaining ship, the *Victoria,* and sailed it west to the Indian Ocean. From there, the ship followed the African coast, making many futile attempts to round the Cape of Good Hope over several weeks.

Del Cano was a sailor who had been promoted to captain and did not have Magellan's navigational skills. He therefore wasted much time en route to Spain, as the courses he set were far from optimal. In early May 1522, the *Victoria* finally rounded the Cape, its crew almost dead from hunger and scurvy. In the Cape Verde Islands they bartered for rice the spices they had brought from the East and were able to relieve their hunger. They continued to the Azores and finally arrived in Spain in early September 1522. Of the more than two hundred people to leave Spain three years earlier, only fifteen feeble and famished men returned. But they had achieved the first circumnavigation of the globe.

Magellan had charts of parts of the world, including some Pacific island chains to which European mariners had sailed before him. The compass courses he set on his voyage were guided by these known locations. His success in navigating huge tracts of open ocean owed much to his skill in setting course by compass to a faraway destination, rather than sheer luck in arriving at unknown locations. After the completion of his voyage by del Cano, a chart of much of the world was drawn by Spanish cartographers, constituting one of the great achievements of the harrowing voyage. Later in the sixteenth century, Sir Francis Drake brought with him a copy of the new world chart and used it in his own sailing.

In the wake of Magellan's achievement, the confrontation between Spain and Portugal intensified, each claiming sovereignty over islands and locations in the Indian Ocean and elsewhere. Negotiations between the two seafaring nations continued, however, and one result of this communication was that Spain was able to recruit the services of a number of excellent Portuguese cartographers. This allowed the Spanish government to improve the navigational knowledge of Spanish mariners, resulting in more efficient voyages across the oceans. Del Cano himself commanded a ship in a fleet sent in 1525 to replicate Magellan's circumnavigation. He died at sea, and most of the ships of his expedition were lost. In 1529, with the treaty of Saragossa, Spain sold Portugal the rights to the Moluccas, and further boundaries between areas under Spanish and Portuguese influences were established.

The arrival of the *Victoria* in Spain marked the greatest achievement of the Great Age of Exploration. During this period, all the world's major oceans became known to Western navigators, and the fact that the world is round was firmly established by Magellan's circumnavigation. The navigators of the late fifteenth and early sixteenth centuries also demonstrated that there is no ocean large enough that it cannot be crossed with good accuracy of navigation using a compass, an astrolabe to measure stars' heights above the horizon, and little more.

In the seventeenth century, Dutch navigators sailing south from Cape Horn discovered Australia, and in the eighteenth century, Bering discovered the Pacific passage to the Arctic Ocean. The following century, Captain James Cook sailed around New Zealand, discovered Hawaii, and searched for a passage from the Pacific to the Atlantic Ocean through Alaska.

Captain Cook was the last great navigator to contribute to our understanding of the workings of the magnetic compass, as well as to benefit immensely from the advantages of the compass in his voyages of exploration. Cook studied the variation of the magnetic compass in a scientific way. He made extensive measurements of the magnetic declination in the areas in which he sailed (by comparing compass readings with computed results from astronomical observations). His work resulted in accurate chart markings of the magnetic variation around the earth. Captain Cook's great achievements in navigation mark the pinnacle of the success of the magnetic compass.

TWELVE

———◦(◦)◦———

Conclusion

I LOOKED UP FROM THE DUSTY OLD VOLUMES STREWN all over the large table in front of me at the Center for the Culture and History of Amalfi. I'd been sitting at the center for many hours, and I was tired and my vision blurred. But in my mind's eye, I was beginning to see clearly the role the compass played in the history of the world. The ingenious invention was finally yielding its secrets.

———————

The story of the magnetic compass demonstrates that the right invention at the right time can change the world. A great invention can lie dormant or be used for secondary purposes for a very long time and then suddenly be discovered by the right people—individuals with vision and an entrepreneurial spirit—and be exploited to its fullest extent. When this happens, such inventions can change the way we live.

The compass was invented in antiquity in China, where it did not immediately improve navigation but was used in feng shui. Both the compass and gun powder—the two greatest Chinese inventions—were exploited to their fullest potential not by their Chinese inventors but by Europeans: the former for a productive purpose, the latter destructive. China may not have been the place where an invention such as the compass could be fully developed or used, or its knowledge disseminated to others. A modern example supporting this assertion is the story of the fight against malaria. In recent years, quinine has lost its potency because the parasite that causes malaria developed a resistance to this traditional cure. In China, however, an herbal medicine for malaria has been known for centuries. This discovery, just like that of the compass, was kept secret. Only in the 1990s did the West obtain enough information from obscure Chinese sources to be able to identify the chemical composition of the agent. As it turned out, the plant from which the drug can be derived grows wild in the United States and other Western countries. Finally, the worldwide battle against malaria was given a fighting chance.

Once the idea of the magnetic compass became widely known, at the end of the twelfth century, the ground was broken for this invention to be implemented in navigation, where it could produce the greatest benefit. Fortuitously, at that time, there was a maritime power in Europe that was able to put the compass to use—and to improve it to the point where it could be employed efficiently in navigation to indicate all directions, not just north and south. That maritime city-state was Amalfi, and during its brief moment on the world stage when it could make a difference, it did.

But soon, powers changed and the Venetians, with their celebrated fleet, were the first to truly exploit the new, improved magnetic compass, thereby bringing navigation in the Mediterranean to a new level. With their great shipbuilding facility, the Arsenal, the Venetians had the capability to construct large ships; the new invention, the compass, made the shipbuilding technology useful. Large ships such as the *Roccaforte* would have been of little use if they could not be sailed in winter and if they could not be sailed accurately.

The technological revolution that created the compass also led the way to charts and pilot books, and with these developments came large ships, frequent voyages, and the resulting rise in prosperity. Venice became the Queen of the Sea in large part by exploiting an ancient idea and using it to meet modern needs.

The next stage of world development came with the Great Age of Exploration, when Columbus, da Gama, Magellan, and other Spanish and Portuguese navigators conquered the oceans and opened up new trade routes to places that were not accessible before their voyages. Here, the magnetic compass found its most consistent, and often exclusive, use as a navigational instrument. Charts of the Atlantic and Pacific Oceans were generally not available to these courageous mariners. The depth of the sea was unknown, and there was little knowledge of shores and islands and inlets. In the vastness of the ocean, a captain had to rely on the floating wind rose of the magnetic compass and on celestial observations.

The compass allowed mariners to chart the oceans and to establish sea routes traversing the entire globe. We use the same sea routes today, and they connect the world's economies to one another. Ships sailing across the Pacific, laden with thousands of products bound from the East, still use a compass not too different from Magellan's, even if today it is mainly powered by electricity (the gyrocompass). We are hardly aware of this fact, or how the compass connects the world, even though in our daily lives we encounter and use so many products made in China and other faraway lands and shipped to us from across the ocean.

The Riddle of the Compass

The world had to wait many centuries for the invention of the compass to take hold and to be applied to navigation. But the story of how a technology has to wait repeats itself again and again. Thirty-five years ago, while sailing with my father across the Atlantic on the SS *Theodor Herzl*, our ship ran into a hurricane. As we neared the eye of the storm, the wind grew stronger and the waves higher. But my father had a wonderful technological tool at his disposal, which helped him avoid the worst part of the storm. On the wall of his chart room, there was a gray machine, and when my father pressed a button, a transmitted weather report began to unroll. A bluish sheet of paper—a chart made of many dots forming curves and numbers, indicating the storm's estimated location and intensity—was slowly appearing before our eyes. The machine that produced this up-to-the-minute weather forecast was the first model of what we know today as a fax machine. For years these machines were used exclusively for transmitting weather maps to mariners and pilots. Only relatively recently did the invention become so prevalent commercially. I can still remember the sensation the fax caused when it first became so popular ("Can you believe people have restaurants fax them their menus to the office?").

The copy machine, the Internet, color television, and cellular phones all could have been popularized decades before they were. The technology to produce and implement these inventions has been in existence for years. The Internet began as a network of interconnected computers used by university researchers and the military in the 1960s. Mobile phones have been used by a few individuals for as long; and the copy machine had its origins very early in the twentieth century. Color television was invented in 1929. The list goes on and on. It seems to be a law of nature that a technology is developed and then waits a long time for people to discover their need for it, rather than the other way around. The time and place have to be right for the implementation of a new technology—but once the conditions are right, the technology can change our lives.

The magnetic compass was the first technological invention after the wheel to change the world. From its origins in ancient China, through the Middle Ages, and on to our time, the compass has been used and improved. Today, electronic compasses are still the most important navigational tool in use on ships and airplanes. Of course, the Global Positioning System (GPS), which uses satellites, has replaced celestial observations by sextant.

"So you have finished," said the archivist with his gentle smile. "Yes," I answered, looking up at him, rubbing my eyes, "but I still don't know whether or not there was a Flavio Gioia, inventor of the compass." "It all depends on a missing comma," he said knowingly, and I was sure that he had read every word in the many ancient volumes in his care. "Well, Professore," he added, "from here, you continue on your own. Good luck to you." I rose and shook his hand. I thanked him for everything he had done for me during my stay in Amalfi. *I'll miss him,* I thought. Then I went down to the main square.

I stopped in front of the bronze statue. The pedestal was fringed with beautiful flowers. *They really do admire him here,* I thought, *whoever he might have been.* A bus carrying tourists arrived, and a group of the travelers crowded around the statue. They stood there for a moment, trying to decipher the Italian inscription. "This guy invented the compass," one of them said as they all walked away. I looked up at the statue and thought: *Flavio Gioia, if you ever did exist, you have no idea what an impact your invention has had on the world.*

A Note on the Sources

By its nature, my research of the questions about the origins of the compass entailed a detailed study of many arcane sources. These were books, manuscripts, and pamphlets of specialized academic circulation, and many were not readily available in libraries. In addition, many of the sources were not in English. A significant number of my references were manuscripts and papers written in Europe and China hundreds of years ago; and even the recent sources, scholarly books written in the nineteenth and twentieth centuries, were often written in Italian, French, or German. Translating the source material has been both challenging and rewarding for me and has made the research for this book the most interesting project I have ever undertaken.

However, because of the large number of sources used and the fact that these materials are not accessible to most readers, I have refrained from making frequent references to the sources within the text itself. To do so would have entailed inserting several footnotes on every page of the book, which would have interrupted the flow of the narrative. In lieu of references within the text, what follows is a summary of the more important sources used in each chapter. The

A Note on the Sources

interested reader may look up any of these references (listed by author and year) in the references section following this note.

CHAPTER 1: The two-hundred-year-old treatise written in French and published in Naples is Venanson (1808).

CHAPTER 2: Much of the material on pre-compass navigation is from Taylor (1956). The information on astronomical observations in antiquity is from Neugebauer (1952). See Walker (1997) for a magnetic sense in animals. The report on the recent discovery of the shipwreck is from an article by William J. Broad, *The New York Times,* Tuesday, March 27, 2001.

CHAPTER 3: The European compass is discussed in Neckam (1187), Provins (1208), Vitry (1220), Peregrinus (1269; trans. 1902), Chaucer (1892), May (1955), Marcus (1956), Taylor (1956), White (1962), and Kreutz (1973). The translations of Dante are from Allen Mandelbaum's wonderful books, *The Divine Comedy of Dante Alighieri* (University of California Press, 1988). Translations of other Italian poems are my own.

CHAPTER 4: The references to the Tower of the Winds and the wind rose are from Motzo (1947) and Kreutz (1973). Other information on the history of navigation is from Taylor (1956).

CHAPTER 5: References to the Amalfi compass are from Pansa (1724), Bertelli (1901), Proto-Pisani (1901), Porena (1902), Apuzzo (1964), and Gargano (1994).

CHAPTER 6: A key reference on the origins of the Italian compass is Mazzella (1570). Also important are Proto-Pisani (1901), Porena (1902),

A Note on the Sources

and Gargano (1994) and references therein. Also a reference to Flavio Gioia is in Nuce (1668).

CHAPTER 7: The information on the Chinese invention of the compass is from Gaubil (1732), Tseng (1935), Wang (1949), and mostly Li Shu-Hua (1954) and Needham (1962).

CHAPTER 8: Much of the information on the history of Venice is from Lane (1973), Norwich (1982), and references therein.

CHAPTER 9: Information on Marco Polo is from his book (1298) and Parry (1974).

CHAPTER 10: The compass and associated aids such as early charts and pilot books are discussed at length in Motzo (1947). Also important is Lane (1963).

CHAPTER 11: References to navigation with the compass are from Morison (1942), Marcus (1956), Taylor (1956), Lane (1963), and Stimson (1990). See also Parry (1974).

References

Alighieri, Dante. *The Divine Comedy*. Translated by Allen
    Mandelbaum. Berkeley: University of California Press, 1988.

Al-Kibjaki, Bailak. *The Book of the Merchants' Treasure*. Cairo, 1282.

Apuzzo, Aniello. *L'Invenzione della bussola e Flavio Gioia*. Naples:
    Rinascita Artistica, 1964.

Barberino, Francesco da. *I documenti d'amore*. Florence, 1318.

Bertelli, P. Timoteo. "Sull'anniversario della bussola." *Corriere di
    Napoli*, 22 May 1901.

———. *Discussione della legenda di Flavio Gioia, inventore della bussola*.
    Pavia, 1901.

Broad, William J., "In an Ancient Wreck, Clues to Seafaring Lives."
    The New York Times, Tuesday, March 27, 2001.

Brown, Charles H. *Nicholl's Concise Guide to Navigation*. Glasgow:
    Brown, Son and Ferguson, 1989.

Casson, Lionel. "The *Isis* and Her Voyages." *Transactions of the
    American Philological Association* 81 (1950): 43–48.

Chaucer, Geoffrey. *The Complete Works of Geoffrey Chaucer*. Edited by
    Walter W. Skeat. London, 1892.

Gargano, Giuseppe. "Fortificazioni e marineria in Amalfi Angioina."
    Rassegna del Centro di Cultura e Storia Amalfitana 14 (December
    1994): 101–3.

References

Gaubil, Antoine. *Observations mathématiques, astronomiques, géographiques, et physiques tirées des anciens livres Chinois.* Paris: Rollin, 1732.

Homer. *The Odyssey.* Translated by Robert Fitzgerald. New York: Random House, 1961.

Hourani, George Faldo. *Arab Seafaring in the Indian Ocean in Ancient and Early Medieval Times.* Princeton Oriental Studies, no. 13. Princeton, N.J.: Princeton University Press, 1951.

Kreutz, Barbara. "Mediterranean Contributions to the Medieval Mariner's Compass." *Technology and Culture* 14, no. 3 (July 1973): 367–83.

Lane, Frederic C. "The Economic Meaning of the Invention of the Compass." *The American Historical Review* 68, no. 3 (April 1963): 605–17.

———. *Venice: A Maritime Republic.* Baltimore: Johns Hopkins University Press, 1973.

Leisegang, Hans. "The Mystery of the Serpent." In *The Mysteries: Papers from the Eranos Yearbooks.* Bollingen Series 30. New York: Pantheon, 1955.

Lipenico, V. Martino. *Navigatio Salomonis Ophirica.* Frankfurt, 1660.

Marcus, G. J. "The Mariner's Compass: Its Influence upon Navigation in the Later Middle Ages." *History* 61, no. 1 (1956): 16–24.

May, W. E. "Alexander Neckam and the Pivoted Compass Needle." *Journal of the Institute of Navigation* 8 (July 1955): 283–4.

Mazzella, Scipione. *Descrittione del Regno di Napoli.* Naples, 1570.

Meilink-Roelofsz, M. A. P. *Asian Trade and European Influence in the Indonesian Archipelago between 1500 and 1630.* The Hague: Nijhoff, 1962.

Morison, Samuel Eliot. *Admiral of the Ocean Sea: A Life of Christopher Columbus.* Boston: Little, Brown, 1942.

References

Motzo, B., ed. *Il Compasso da navigare*. Cagliari: University of Cagliari, 1947.

Needham, Joseph, F.R.S. *Science and Civilisation in China*. Volume 4, part 1, *Physics*. Cambridge: Cambridge University Press, 1962.

Neckam, Alexander. *De naturis rerum*. London, 1187.

Neugebauer, O. *The Exact Sciences in Antiquity*. Princeton, N.J.: Princeton University Press, 1952.

Norie, J. W. *Norie's Nautical Tables*. London: Imray, Laurie, Norie and Wilson, 1941.

Norwich, John J. *A History of Venice*. New York: Knopf, 1982.

Nuce, D. Angelus de. *Neapolitanus*. Paris, 1668.

Pansa, F. M., ed. *Istoria*. Naples, 1724.

Parry, J. H. *The Discovery of the Sea*. New York: Dial Press, 1974.

Peregrinus, Peter. *Epistle to Suggerus of Foncaucourt, Soldier, Concerning the Magnet*, 1269. English translation: London, 1902.

Polo, Marco. *The Travels*. 1298. English translation by Ronald Latham. New York: Penguin, 1996.

Porena, Filippo. *Flavio Gioia: inventore della bussola moderna*. Rome: Direzione della Nuova Antologia, 1902.

Proto-Pisani, Nicolangelo. *Sull'origine della bussola*. 1901. Reprint, Salerno: Libreria Antiquaria Editrice, 1973.

Provins, Guyot de. *La Bible*. Cluny, 1208.

Schück, Albert. *Der Kompass*. 3 Volumes. Hamburg: Selbsverlag des Verfassers, 1911–18.

Shu-Hua, Li. "Origine de la boussole." *Isis* 45 (1954): 175–96.

Sobel, Dava. *Longitude*. New York: Walker, 1995.

Stimson, Alan. "The Longitude Problem: The Navigator's Story." In *Quest for Longitude,* edited by William J. H. Andrewes. Cambridge, Mass.: Collection of Historical Scientific Instruments, Harvard University, 1996.

References

Taylor, Eva G. R. *The Haven-Finding Art: A History of Navigation from Odysseus to Captain Cook.* London: Hollis and Carter, 1956.

Tseng, K. L. *Collection dans le K'in-ting.* Shanghai: Commercial Press, 1935.

Venanson, Flaminius. *De l'invention de la boussole nautique.* Naples, 1808.

Vitry, Jacques de. *Historiae Hierosolymitanae.* Paris, 1220.

Walker, Michael, et al. "Structure and Function of the Vertebrate Magnetic Sense." *Nature* 390 (1997): 371–6.

Wang, T. "Aiguille montre-sud." *Chinese Journal of Archaeology* 4 (1949).

White, Lynn, Jr. *Medieval Technology and Social Change.* New York: Oxford University Press, 1962.

Winter, Heinrich. "Who Invented the Compass?" *Mariner's Mirror* 23, no. 1 (1937): 95–102.

Acknowledgments

—————⋙《❂》⋘—————

I thank my editor and friend, Jane Isay of Harcourt Inc., for together with me coming up with the wonderful idea of a book about the compass. I am also grateful to her for her patience throughout the long period it took me to complete the book and for her encouragement and support. I thank Jennifer Aziz for her tireless efforts to obtain the figures and pictures used in the book and for her help in the preparation of the manuscript. I thank Rachel Myers for her careful editing of the manuscript and for many suggestions and corrections. I thank David Hough for his advice and suggestions and for managing the production of the book.

I thank Professor Daniel Ruberman of Brandeis University for arranging for me to be a visiting scholar at Brandeis during the period this book was being researched and written. I thank the librarians at Brandeis University and Bentley College for helping me obtain rare books needed for the completion of this project.

I am most grateful to Mr. Giuseppe Cobalto of the Center for the Culture and History of Amalfi for providing me access to a large number of very important sources on the history of the compass and for his generosity and patience in answering many questions. I found

Acknowledgments

the periodical published by the Center in Amalfi, the *Rassegna del Centro di Cultura e Storia Amalfitana,* edited by Mr. Cobalto and his colleagues, useful in the preparation of this book.

I thank the curator of the National Maritime Museum in Haifa for permission to reprint a picture of a Chinese jade disk used in navigation. I thank Captain Hillel Yarkoni for help in expediting the permission.

I thank Dr. Paolo Bruschetti, the director of the Museo dell'Accademia Etrusca in Cortona, Tuscany, for kind permission to reprint the image of the Etruscan *Lampadario di Bronzo* and for an interesting interview about the artifact.

I thank my wife, Debra, for all her help and encouragement. This book is dedicated to her.

Index

Adriatic Sea, 97–98, 99, 127
Africa, 135–36, 138–41
Alaric I, 93
Alexander the Great, 111
Alexander VI, 136
Alexandria, 130–31
Alexius, 98
Alexius I, 58
Alfonso I, 55–56
Alfonso of Aragon, 62
Amalfi, 53–75, 98, 155
    Center for the Culture and
      History of Amalfi, 6–7, 153,
      158–59
    founding of, 57–60
    Flavio Gioia and, 5–7, 68–75,
      158–59
    Naples and, 53–57
    origins of compass and, 4–7,
      68–75, 153, 158–59
Andronicus of Macedonia, 40–41
Apeliotes, 42, 43–44
Argestes, 42

Argonauts, 51
Arsenal, in Venice, 100, 155
Arsinoeion, 50–52
astrolabe, 136
astronomy, 18–28
    celestial bodies in, 18–19, 22–23,
      138, 140, 145
    Chinese, 23–25
    Egyptian, 18–21
    Greek, 18–19, 22–23
    latitude measures and, 19–20,
      21
    longitude measures and, 20
    Polaris (North Star) and, 22–23,
      118
    Portuguese, 136
    Roman, 21–22
Atlantic Ocean, 13, 134, 136, 137,
    140–41, 156
Attila the Hun, 93
augurs, 46–50
Australia, 150
Azores, 148

Index

Babylon, 18–19
Bailak, 120–21
Baldwin of Flanders, 103
Baltic Sea, 134
barbarian invasions, 92–95, 97, 98,
    113, 116–17
Barberino, Francesco da, 33–34,
    36–37
Bay of Naples, 54–55, 59–60
Beccadelli, Antonio, 61
Bertelli, Timoteo, 6–7, 65, 66–71, 72,
    75
Bible, 9–11, 39–40
Big Dipper, 22–23, 78–79, 80, 90
Biondo, Flavio, 62, 66–71
bird migration, navigation and,
    16–18
Black Death, 60, 105
Black Sea, 112, 115
Bonaventure, Saint, 35
Boreas, 42, 43–44
Bourbons, 56
boxed compass, 36, 61, 75
Brazil, 145
Bronze Age, 10
bubonic plague, 60, 105
bussola, 5, 36, 71, 75, 124
Buti, Francesco da, 36
Byzantium, 92, 95–96, 98, 99

Cabral, Pedro Alvarez, 143
calendars, 18–19
Calypso, 25
Cano, Sebastián del, 148, 149
Cape Comorin, 118

Cape Horn, 141, 150
Cape Malea, 14–15
Cape of Good Hope, 138–39,
    141–42, 148
capes, 14–15
Cape Verde Islands, 140, 141, 145, 148
Caro, Lucrezio, 66
carta, 124
Carta Pisana, 61–62, 125–26
cartography, 128–29, 149
Chares of Lindos, 15–16
Charlemagne, 95
Charles I of Anjou, 55, 71–72
Charles II of Anjou, 72
charts, nautical, 13, 44–45, 61–62,
    123, 124–31, 134, 149, 156
Chinese civilization
  astronomy, 23–25
  Church and, 89–90
  dry-pivoted wooden turtle, 84,
    85, 86, 88
  herbal medicine, 154–55
  iron-fish floating compass, 81–83,
    86, 88, 121
  jade disk navigation, 23–25
  navigation methods, 86–87,
    114–15, 117–18
  origins of compass in, xiii, 65,
    77–90, 114–15, 120–22, 154
  south-controlling ladles and
    spoons, 78–80, 81–83, 86, 87
  trade with, 111–17, 119–20, 142
  water-floated magnetic needles,
    84–85, 86, 121
Ch'in Shi Huang Ti, 78

Index

chronometers, 136

Clemens, Samuel, 12

color television, 158

Colossus of Rhodes, 15–16

Columbus, Christopher, 136–37, 144, 156

compass

as divining tool, 88–90

improvements in overall design of, 73

literary references to, 29–37, 62, 65, 66, 67, 69, 78–83, 120–21

mystical origins of, 45–52

nature of, xiv

operation of, xiii–xv

origins in China, xiii, 65, 77–90, 114–15, 120–22, 154

origins in Europe, xiii, 4–7, 53, 68–75, 118–19, 153, 158–59

poles of, xiv

Marco Polo and, 114–15, 118–22

reliability of, xvi

significance of, xii–xiii

Venetian trade and, 103–7, 114–15

compass cards, 34–35, 36–37, 61

Compasso, 124, 129

Compasso, Bartolomeo, 71–72

Constantine the Great, 57

Constantinople, 55, 56–58, 92, 102–3, 112, 115, 120

Cook, James, 150–51

Coppa Tarantina, 50

core of earth, xiv

Corfu, 58, 130–31

Crete, 10, 15, 106, 124, 130–31

Crusades, 31, 98, 99, 102–3, 107, 108, 112–13

Cumae, 54

currents, navigation and, 16

Cyprus, 130

da Gama, Vasco, 139–44, 156

Dante Alighieri, 32–33, 34–36, 68, 100

Dati, Leonardo, 34

dead reckoning, 129–31, 137

Dee, John, 33

depth markings, 13

Dias, Bartholomeu, 141

divination, 51–52, 88–90

Divine Comedy, The (Dante), 32–33, 34–36, 100

Drake, Francis, 147, 149

dry pivoted compass, 33, 84, 85, 86, 88

dynamo action, xiv

Earth

magnetic fields of, xiii–xv, 17–18

rotation about axis, 22

shape of, 21–22

Eastern Roman Empire, 55, 56, 57–58, 92, 95–96, 97

ecliptic, 23–25

Egyptian civilization, 10, 18–21, 42–44, 50, 56, 57

eight-wind system, 40–41, 43, 45, 47

Etruscan chandelier, 48, 49–50

Etruscan civilization, 45–50, 52

Index

fax machines, 157
feng shui, 87–89, 118
floating compass, 33, 81–83, 86, 88, 121
Franks, 95–96, 97
Frederick II of Hohenstaufen, 55

Gaeta, 56
Gargano, Giuseppe, 70
Genoa, 58, 61, 104, 106, 107, 112, ·130–31, 143
Ghenghis Khan, 113
Gilbert of Colchester, 72
Gioia, Flavio, 5–7, 63–75, 158–59
Giri, Bernardo, 72
Global Positioning System (GPS), 136, 158
Godfrey of Bouillon, 99
Goths, 55
Goya, Roberto de, 72
Great Age of Exploration, 107–9, 133–51
  Christopher Columbus, 136–37, 144, 156
  James Cook, 150–51
  Vasco da Gama, 139–44, 156
  Francis Drake, 147, 149
  Henry the Navigator, 135–37, 138
  Ferdinand Magellan, 144–50, 156
  Portuguese in, 135–37, 138–50, 156
  Spanish in, 136, 144–50, 156
Great Khan, 115, 116–17
Great Wall of China, 88

Greek civilization, 9–10, 14–15, 18–19, 22–23, 40–41, 46, 50, 53–54, 99
Guam, 147
Guinizelli, Guido, 32–33
Guiscard, Robert, 57–58, 98
Guyot de Provins, 30–31
gyrocompass, 156

Hapsburgs, 56
Hawaiian Islands, 11, 150
Helios, 15–16
Henry the Navigator, 135–37, 138
Homer, 14–15
Horn of Africa, 141, 150
Hsuan-Chi (jade disk), 23–25
Huns, 93

Iceland, 16–17, 133–34, 135
India
  coast of, 16
  trade with, 138, 142, 143, 144
Indian Ocean, 13, 111, 117, 121, 137–39, 142, 148, 149
Indicopleustes, Cosmas, 111
Internet, 157–58
Ioha, Francisco de, 72
Irish monks, 16
Iron Age, 78
Israel, 10–11, 39–40
Istria, 95

jade disks, 23–25
Java, 144

Index

Jesuits, 89
Jonah, 10–11

Ibn-Khaldūn, 129
Khou Tsung-Shih, 121
Kochab (Beta Ursae Minoris), 22–23
Kublai Khan, 116

Large Magellanic Clouds, 145
latitude, 19–20, 21, 145
lighthouses, 15
Li Shu-Hua, 80–83, 120–22
Little Dipper, 22–23
Li Ying-Shih, 89
lodestone, xii–xiii, xiv, 31, 51, 78–79, 80, 84–86, 129
Lombards, 93–94, 95
longitude, 20, 136, 145
Lopez de Sequiera, Diego, 143–44
Louis, Saint, 121

Macao, 144
Ibn-Mādjid, Ahmad, 142
Magellan, Ferdinand, 144–50, 156
magnetic fields, xiii–xv, 17–18
Malacca, 143–44
Malamocco, 95, 96
malaria, 154–55
Malta, 11, 13
Manoel, King, 142
mappamundi, 129
Marianas, 148
mariners, defined, 101. See also Great Age of Exploration

Marquesas Islands, 11
Mauro, Fra, 134
Mazzella, Scipione, 69, 70
Mediterranean Sea, xiii, 10, 11, 13, 21, 36–37, 39–40, 43, 50, 52, 54–55, 58–59, 78, 97, 101, 112, 123–31, 134
Menelaus, 14–15
Mèng T'ien, 88
Middle Ages, 36–37, 47, 57, 61, 104–5, 123, 158
migration patterns, navigation and, 16–18
Minoan civilization, 10
Mississippi River, 12
mobile phones, 158
Moluccas, 148, 149–50
Mongols, 113, 116–17, 122
monsoon winds, 137–38
Morocco, 140
Moslems, 97, 112–13, 119–20, 136
Mosto, Alvise Ca'da, 125
Motzo, Bacchisio, 47–49
Murano, 94, 101
mystai, 50

Naples, 53–57, 58, 62, 64
Napoleon Bonaparte, 109
Natal, 142
nautical charts, 13, 44–45, 61 62, 123, 124–31, 134, 149, 156
navigation methods
ancient ships in open waters, 9–12

Index

navigation methods (*continued*)
  astronomy, 18–28, 118, 136
  birds and animals, 16–18
  currents, 16
  dead reckoning, 129–31, 137
  lighthouses, 15
  nautical charts, 13, 44–45, 61–62,
    123, 124–31, 134, 149, 156
  pilot books, 123–24, 125, 131, 142
  shore profiles, 14–16
  sounding line, 12–13, 134–35
  sun positions, 23–25, 41
  tides, 13–14
  winds, 16, 130, 137–38, 140–41
  written sailing instructions,
    44–45, 134–35
Naxos, 106
Neckam, Alexander, 29–30, 114, 121
Needham, Joseph, 77–78, 79
New World, 136–37
New Zealand, 17, 150
Nile River, 18
Normans, 57–60, 98
Norsemen, 16–17, 133–34
North Pole
  celestial, 22–24
  geographic, xvi
  magnetic, xiv–xv, xvi, 85–86
North Sea, 134
North Star (Polaris), 22–23, 118
Notus, 42, 43–44

Odyssey, 7, 14–15, 25
Orphic cults, 46–50

Pacific Ocean, 147, 156
Patagonia, 145
Paul, Saint, 11, 13
Peking, 115, 116
Pepin, 95–96
Peter the Pilgrim (Petrus
  Peregrinus), 33
Philippines, 148
Phoenician civilization, 10–11,
  22–23, 46
pilot books, 123–24, 125, 131, 142
Pio, Giambattista, 66–69, 70, 73
Pisa, 58, 60, 104, 112, 125–26, 130
Pizzigani, Francesco, 128–29
Pizzigani, Marco, 128–29
Pliny the Elder, 55
Pliny the Younger, 21–22, 54–55,
  92
Polaris (North Star) and, 22–23, 118
poles
  of compass, xiv–xv, xvi, 85–86
  of Earth, xiv–xv
Polo, Maffeo, 115–17, 119
Polo, Marco, 107, 113–22, 144
Polo, Nicolò, 115–17, 119
Polynesia, 11
portolano, 124, 125, 126–27
Portugal, 135–37, 138–50, 156
Prague disk, 51, 52
precession, 22, 23
Ptolemy II, 42–43

Ravenna, 95
ravens, 16–17

Index

Red Sea, 19, 20, 21, 40
Rhodes, 15–16, 130
Robert of Anjou, 72
Roccaforte (ship), 105–6, 155
Roger II, 55
Roman civilization, 11, 21–22, 45,
    46–48, 54–55, 56, 92–93,
    111–12, 119–20, 130

Saladin, 113
salmon, 17
Samothrace, 50–52, 120
Santorini, 10
Saracens, 97
sextant, 136
Shen Kua, 84–86
shore profiles, 14–16
Sierra Leone, 140
silk road, 111–12
Sirius, 18, 20–21
sixteen-wind system, 36–37, 45,
    47–52
Small Magellanic Clouds, 145
Sobel, Dava, 20
Solomon, King, 10–11, 40
sounding line, 12–13, 134–35
South Sea, 147
Spain, 136, 144–50, 156
spice trade, 143
Stadiasmus of the Great Sea, 124
Strait of Magellan, 145–47
Strait of Messina, 3–4
sun, positions of, 23–25, 41
Syria, 21, 56, 57, 97, 99, 121

Taoist philosophy, 87–88, 90
Tartars, 113
Taylor, E. G. R., 41–42
television, 158
Thales, 22–23
Theophylactus, 112
thermoremanence, 81
tides, navigation and, 13–14
Timosthenes, Aristotle, 42–44
Tower of Winds, Athens, 40–41, 42
trade
    Great Age of Exploration and,
        138–44
    Venetian, 97–105, 112, 143, 155
Tsêng Kung Liang, 81
Twain, Mark, 12
twelve-wind system, 36–37, 41–44,
    47, 51–52

Varthema, Lodovico di, 144
Venice, 56–59, 61, 91–109
    barbarian invasions, 92–95, 97, 98
    Crusades and, 98, 99, 102–3
    Great Age of Exploration and,
        107–9
    magnetic compass navigation
        and, 103–7
    Marco Polo and, 113–17
    shipbuilding in, 99–101, 105–6,
        155–56
    as trading center, 97–105, 112,
        143, 155
    year-round sailing, 130–31
Vesconte, Petrus, 127–29

Index

Vespucci, Amerigo, 144
Vesuvius, 54–55
Victoria (ship), 148, 150
Vikings, 16–17
Virgil, 35
Visigoths, 93
Vitry, Jacques de, 31

Walker, Michael, 17–18
Wang Ch'ung, 80
Wang Mang, 79–80
weather maps, 157
wind rose, 36–37, 40, 42–43, 61, 123,
    124
winds
    eight-wind system, 40–41, 43, 45,
        47

navigation using, 16, 130, 137–38,
    140–41
sixteen-wind system, 36–37, 45,
    47–52
twelve-wind system, 36–37,
    41–44, 47, 51–52
Wu Ching Tsung Yao, 81–83, 90, 121

year-round sailing, 130–31

Zephyr, 42, 43–44
Zeus, 14–15
zodiac, 19